D0908292

THIS SCULPTURED EARTH

John A. Shimer

This SCULPTURED EARTH:

THE LANDSCAPE OF AMERICA

NEW YORK · COLUMBIA UNIVERSITY PRESS · 1959

Acknowledgments

I<small>N</small> telling the geologic story of the varied landscapes of the United States I have been most fortunate in having a number of people around me whose interest and enthusiasm have been a tremendous help. Much of the initial impetus in writing the book was provided by Dorothy Abbe, whose insistent questions about the scenery of our land were most stimulating and needed answering. In addition to being an undeniable catalyst she is also the designer of the book.

Maps and diagrams have been provided by my wife, Genevieve Shimer, as well as much assistance in making the scientific explanations less ponderous than they might otherwise have been. In this latter phase she was joined by Florence Henry Shimer, whose help was wholehearted and very welcome.

During the final stages of work on the manuscript I was able to make a number of improvements following the excellent editorial suggestions of Edwin N. Iino of the Columbia University Press.

J. A . S.

Brooklyn College
May, 1959

Contents

Illustrations

THIS SCULPTURED EARTH

CHAPTER 1

SCENERY means many things to many people. Each part of this sculptured earth has its own characteristic flavor and its own special type of landscape, and each arouses unexpected and varied reactions in the observer.

A map of the United States tells us that the country is composed simply of plains and mountains and plateaus, all enclosed by an irregular shoreline. This is the bare outline. What actually exists is a landscape of prac-

The Meaning of Scenery

tically unlimited diversity. Think of the smooth grass-covered, tawny hills of California, and compare them with the jagged gray peaks of the Tetons, with touches of snow still clinging to them even on the hottest summer days. Picture the myriad pinnacles of Bryce Canyon, glowing in early morning sunlight, and then call to mind the vast monochromatic expanses of the arid desert regions of Nevada. Or imagine the subtle tones of a green carpet of young wheat on the far-reaching plains of the Dakotas, and contrast them with the flaming fall colors of the maples on some intimate hillside in Vermont.

Landscapes are certainly many faceted, and our appreciation of them is generally subjective. A scene is compounded of various elements, the weather, the verdure, the earth itself, and our appreciation depends on how closely we may feel "in tune" with our surroundings at a given moment.

Mountains are the epitome of grandeur, expressive of the awesome power of nature. Among them, a person may perhaps lose most completely his sense of self-importance and acquire a feeling of relative insignificance. For many, such regions offer a challenge and give a wonderful feeling of release and freedom from earthly cares.

The low rolling hills of the eastern Plains States convey an air of quiet serenity. Here the seasons come and go in everlasting cycles, and we are conscious of the passing of time, but not perhaps of dramatic changes in the face of the land such as we can sometimes find elsewhere.

A visit to the seacoast is associated with the eternal activity of moving water. Lazy summer days spent in the hot sun on a warm beach, the tide slowly covering the land, contrast with times of strong wind and heavy surf.

An increased understanding of the various forces which have been important in molding our environment brings an even greater awareness and appreciation of our scenic heritage. Rather than decreasing our feeling of wonder such knowledge often intensifies our reaction. Also with increased understanding the seemingly haphazard arrangement of hills and valleys, plains and mountains, becomes less puzzling and we may even learn to recognize that the infinite number of scenic forms are really variants on a limited number of basic themes.

For a true appreciation of any landscape we must *see* and *understand* the varied shapes of hills and valleys as well as of the smaller scale ledges and slopes, and not be diverted by the manifold impressions created by vegetation, with its variety of shapes, colors, and odors, and the sounds and movement of running water, wind, and waves.

As seen from the air the landscape below us reveals arrangements of features which are difficult to perceive on land and which pose many questions for the traveler. For instance, on a flight from New York City to Quebec we see a striking number of lakes and swamps in an area with a surprising amount of forest cover, considering the large population of New England. In contrast, on a flight over the Appalachians from New York to Pittsburgh we notice a marked paucity of lakes and swamps. On such a flight we might also note a series of parallel forested ridges with intervening cultivated valleys somewhat to the west of Harrisburg, and farther west still nearer to Pittsburgh, an irregular land pattern, one of winding valleys and small irregular fields. In contrast, flying over the plains area of Iowa and Nebraska we see a network of large rectangular fields and straight roads extending to the horizon. Observations such as these are sure to raise questions, many of which can be easily answered by a consideration of the various geological forces at work in the world.

The mountains of Death Valley and the arid areas of Nevada and Utah have a ragged tattered appearance. When we look down at such scenes we are at once aware that a great deal of material has been washed down from the mountains and spread out as alluvial fans in the intervening lower-lying places. It is clear that the ridges and pointed tops of the mountain ranges are residual features, left by the forces that have been carving the valleys. Here, we should probably not talk of hills rising above valleys but of valleys cut between hills. The relationship of hills and

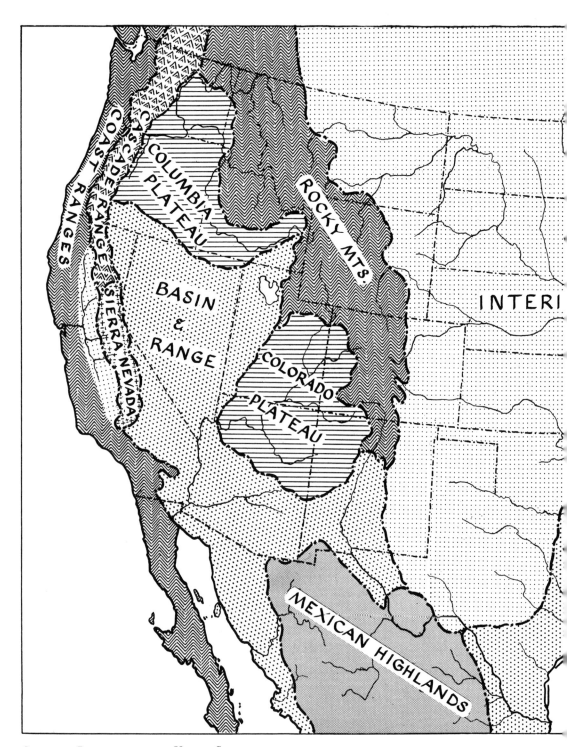

GEOLOGIC PROVINCES OF THE UNITED STATES

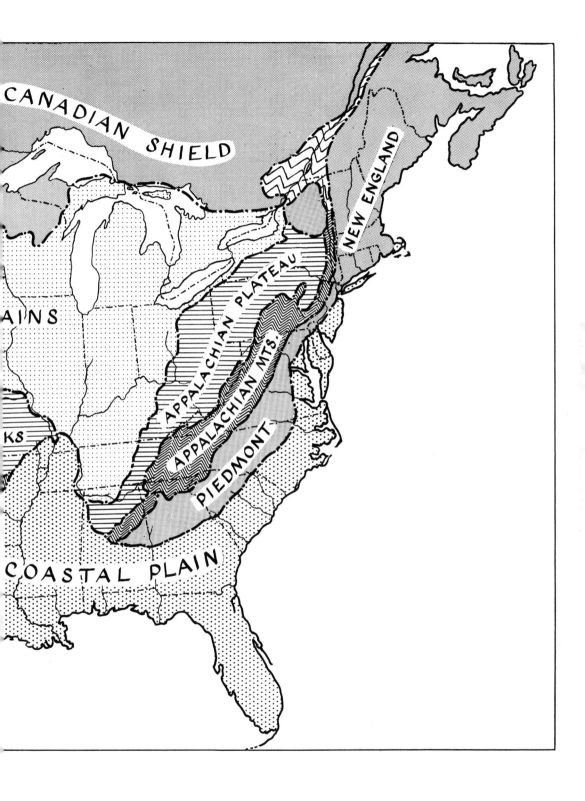

valleys is clearly evident in such an arid area. It is less obvious but equally true in the Appalachians and other mountainous areas where vegetation cloaks and obscures the details of the slopes.

The more we travel around the country and look at scenery, the more we become conscious of the worn, sculptured look of many parts of the land. A great deal of the world's scenery is obviously due to destruction. Solid rocks crumble on being exposed to the weather and are then carried away in this fragmental state, generally by running water, although the wind, waves, glaciers, or underground water may in places be of paramount importance. We will find that this breakup and removal of rocks, the process of gradation, is active everywhere. It is one of the three major processes affecting the earth's crust.

With the continued attack of the forces of gradation the eventual removal of all mountains into the sea is inevitable. It has been estimated

DEATH VALLEY, CALIFORNIA. The rocks of the mountains have been weathered and the broken material washed into the intermontane basins. The watercourses are marked by salt deposits. *Spence Air Photos, Los Angeles*

that if the present rate of weathering and erosion continues, the high parts of North America will be washed into the sea in less than twenty-five million years. We know that the earth is much older than this. The implication is inevitable, therefore, that there must be some counterforces which cause the building up of land areas.

Diastrophism and igneous activity, the second and third major earth-shaping processes, together provide these necessary counterforces. They renew the high parts of the land, which the forces of gradation destroy bit by bit, thus creating as a byproduct the great variety in the scenery of the world.

Igneous activity results in the growth of volcanoes and lava flows. Such features are very obvious at many places in the western part of the United States. The Cascade Mountains of Washington, Oregon, and northern California stand as monuments of the effectiveness of lava and ash in building up the land. The cinder cones of Arizona in the San Francisco Mountain area are beautiful examples of landforms produced by somewhat smaller scale igneous activity. Sunset Crater as seen from the air is an unforgettable sight. It is a practically perfect cone with a small crater at the top; extending from its base there is a black irregularly shaped lava flow which stands out markedly against its lighter toned surroundings.

Evidence of the crustal instability associated with diastrophism is found everywhere in the world. At Pozzuoli on the coast of Italy, for example, the marble columns of the Greek temple of Serapis have small holes drilled in them, fifteen to twenty feet above the present level of the water. These holes were made by boring clams at a time when the columns were covered by the Mediterranean. They bear mute testimony to the drowning and partial reelevation of this part of the world since ancient times.

SUNSET CRATER, ARIZONA. A cinder cone. *National Park Service Photo*

The relationship between land and sea level has been an unstable one from very early times in earth history. Marine shells have been found in very odd places, away from their usual environment along the shore. Fossils of long extinct marine plants and animals are found in layers of limestone, shale, and sandstone, which are the solidified deposits of ancient sea beds, now located far from any present seacoast. Many of these former sea floors are now found at great elevations, in places thousands of feet above the present level of the ocean. During the attempt to scale Mount Everest in 1924, the geologist Noel E. Odell looked for and reported finding marine fossils embedded in sedimentary rock high up on the side of the mountain.

Sedimentary rock layers have often lost their original level position and are now crumpled and contorted. For instance, much of the mass of the Rocky Mountains is composed of such distorted material, originally laid down when that part of North America was under the sea. Now these uplifted sediments are being removed by running water to form new beaches along other oceans. We can frequently see cross sections of some of these ancient tilted and uplifted sea beds as they lie exposed on canyon walls or cliff faces. In Glacier Park, Montana, lines made by such layers are easily noted on the steep slopes of many of the eroded peaks. And along the east margin of the Rockies in Colorado the upturned edges of some of the more resistant sedimentary strata stand out as ridges flanking the mountain front. In the Garden of the Gods at Colorado Springs, the originally horizontal layers are now vertical and stand dramatically as red sandstone walls.

Faulting or slipping along cracks in the crust may occur in places with the result that layers which were once continuous now stop abruptly and continue on the other side of the fault in an offset position. The result of such faulting can be seen especially clearly from the air, as, for example near Loveland, Colorado, where a number of hogback ridges are offset.

The great variety of distortions of the earth's crust is caused by apparently never-ending diastrophic activity, which raises or lowers crustal layers or tilts and crumples them in the process of mountain building.

The human life span is so short that the surface of the earth appears to be substantial. Scenes described long ago seem essentially the same today, and we feel sure will be so for many years to come. Of course, this is an illusion, at least from the geologist's point of view. When a geologist talks about hills which vanish and mountains which rise from the ocean he is obviously using time in terms different from those we commonly use; his time units are in thousands or millions of years. On the

GOING-TO-THE-SUN MOUNTAIN, GLACIER NATIONAL PARK, MONTANA. The steep slopes of the mountain are marked by the horizontal lines of eroded strata.

National Park Service Photo

other hand, a geologist who deals with such a time scale gives great importance to little things which may seem insignificant to others. The washing of an unseeded lawn, the shifting of flagstones by frost action, and the breakup and decay of masonry are the results of exactly the same forces which in the long run can destroy the loftiest mountains. Viewed from the perspective of geologic time the surface of the earth is an ever-oscillating platform, where mountains rise and are washed away and where the sea floods first one part of the land, then another.

In a geological discussion of scenery it is important to appreciate that we are dealing with long periods of time, measured in millions of years, but how many millions in each case is of little importance. Actually it is more important to grasp the sequence of events which have occurred while realizing that we are dealing with time units which are really incomprehensible anyway.

CAPE COD AND MASSACHUSETTS BAY   *Photo by Laurence Lowry*

CHAPTER 2

THE edge of the land where it meets the sea holds a never-ending fasci-
nation for us. Here we are conscious of the timeless quality of waves and
tides, of the eternal ebb and flow of the waters against the land.

Because of the restless motion and turmoil of the ocean, coastal scenery
is one of the least secure, the most evanescent of any on the face of the
earth. We are aware here as nowhere else of the unceasing activity of the

The Edge of the Land

forces that have molded the earth's crust in the past and are continuing to shape it today. Sometimes after a storm slight changes in a familiar pattern are discernible along the seacoast, perhaps a subtle difference in the curve of an inlet or in the outline of a dune against the sky. There may be a new line of pebbles perhaps where there was none before. There are times when storms alter scenery more dramatically, as when waves break through barriers and encroach on the land, making new inlets, new islands, a new shoreline.

The day-to-day or storm-to-storm shifting of the easily moved sand and pebbles along a shore plays a relatively minor part in the long-term action of the sea. Less sudden but extremely impressive is the action of water on the most solid parts of the land, which slowly yield to the ceaseless pounding of waves and the battering of storm-tossed sand and gravel. Steep coastal cliffs which we see today look out over the moving waters that have been carving them for hundreds, perhaps thousands, of years. As an agent of erosion, the sea is powerful indeed. But how a stretch of coast yields to waves, currents, and tides depends of course on the nature of the land encountered by the sea.

Landscapes are created by oceans in yet another, totally different way. A change in sea level produces a unique type of coastal scenery, quite unlike that created by waves. When the level of the sea rises it floods low coastal lands, isolating hills to create islands and drowning river valleys to form bays.

New England affords excellent examples of landscapes along a drowned coast as well as coastal forms resulting from the more direct work of the sea. The scenery along the coast of Maine, in Boston Harbor, and on the outer side of Cape Cod show characteristic results of the sea's activities on contrasting types of land. The ocean drowned the entire coast of New England at approximately the same time. While the rocks along the Maine shore are hard and not easily worn away, the softer materials farther south have yielded readily to the action of the waves and currents.

For the traveler a trip following the edge of the land along the coast of Maine is a tortuous one. Deep estuaries which extend miles inland and form fine harbors alternate with bold headlands jutting well out to sea, promontories from which many islands can be seen filling in the nearer scene before the open sea is reached. The land and sea interpenetrate to a truly remarkable degree. For instance, the distance from Portland, Maine, to the Canadian border is about 200 miles by air, but via the shoreline it is over 2,000 miles—about the distance from Maine to the Rocky Mountains.

From the top of many of the higher hills in the Acadia National Park on Mount Desert Island, an ideal view of this irregular shoreline can be

THE DROWNED COAST OF MAINE. View from the summit of Jordan Mountain, Acadia National Park. The inundation of the sea has made islands and promontories of a former landscape of hills and ridges. The granite in the foreground has been broken into blocks by intense frost action. *Photo by Devereux Butcher*

ROCKS AND WAVES ON THE COAST OF MAINE, at Acadia National Park.

National Park Service Photo

obtained. An island-studded seascape is spread out at our feet. The islands are not haphazardly arranged but appear to be roughly organized into long files, every bit as though they were marching out to sea one after another. There are a number of these island groups in sight, all trending roughly north to south. They seem to be partially submerged continuations of promontories which we can see on the mainland to the north. To the south, toward the horizon the lines of these islands gradually disappear, the underwater ridges getting lower and lower until we can only guess that they continue under water.

A stormy day along this granite coast brings breakers which concentrate their fury on the rocky headlands and outer sides of the islands. Hurling sand and gravel against the land, they have produced such features as Anemone Cave and the neighboring cliffs on the seaward side of Mount Desert Island. Broken material from the headlands has been distributed to form small beaches of sand and gravel between the promontories or sand bars part way across some of the estuaries.

At the same time that the headlands are being worn away, the streams which enter the landward end of the estuaries, miles from the open sea, are bringing sand, mud, and gravel which they dump when their flow is checked on entering the ocean. In some places small growing deltas can be found. They will not be disturbed by the large waves which break on the headlands or by the currents which prevail there. This zigzag, deeply indented coastline of Maine would seem to be heading for inevitable destruction some time in the geologic future. Waves and currents are cutting off and filling in the seaward ends of the bays while landward they are being choked up by stream deposits.

Manifestly this coastline of deep estuaries could not have been produced by the action of waves and currents; explanation of its origin must lie elsewhere. What we see certainly suggests a landscape of hills and valleys which has been submerged, the sea flooding the valleys to form tidal inlets and isolating the hills to form islands, and, to be sure, therein lies the geological explanation. This is a "drowned coastline." It is not surprising therefore that dredging in the harbors along the coast occasionally brings up decomposed grass roots and peat.

THE DROWNED COAST OF MAINE

ROCKLAND

30 MILES

PORTLAND

The remarkable alignment of estuaries and islands is due to the arrangement of the valleys and ridges which existed prior to drowning. The underwater alignment of ledges is the same as that of the islands, and inland, beyond the reach of the sea, the same ridge and valley pattern continues.

The eventual fate of the coast of Maine is inescapable. An essentially straight shore, with purely sea-produced scenery, is inevitable, if the level of the sea does not rise or fall and thus start the cycle of wave erosion all over again. Wave-cut cliffs fronting the sea will alternate with sand and gravel beaches curving slightly back into the land where estuaries were formerly found. But the sea's action on such a rocky coast is slow, and succeeding generations will notice very little change in the contour of the shore.

We see then that the extent of the modifying action of waves and currents on the Maine coast has been slight. A few cliffs have been cut and a few minor beaches and bars have been formed; that is essentially all that the sea has done to the land.

When the sea attacks less resistant material than the rocks of Maine it can produce very noticeable changes in a relatively short time. A comparison of the Maine coast with Boston Harbor demonstrates this difference very nicely. Both areas were drowned virtually at the same time, beginning about ten thousand years ago. The traveler driving around the Boston area is rewarded with a far greater abundance of fine beaches and wave-cut cliffs than are to be found along the coast of Maine, and the impact of the sea on the shoreline is evident almost everywhere.

At Boston the sea has inundated a land liberally covered with piles of loose debris, a great deal of which had been shaped into low rounded hills before the sea rose to its present level. These deposits of gravel, sand,

THE DROWNED COAST AT BOSTON, MASSACHUSETTS. The peninsula of Nantasket Beach is in the foreground. The many elliptical islands are the summits of drumlins. The submergence of the land is also evident in the many estuaries.

Photo by Laurence Lowry

and mud all mixed together, called "drumlins," were left by the large continental ice sheet which disappeared from New England in the very recent geological past, about 10,000 years ago. Most of the drumlins have an elliptical ground plan, and are so oriented that their long axis runs roughly northwest to southeast, reflecting the direction of motion of the glacier as it overrode them and smoothed them into their present stream-lined shape. They vary somewhat in size, commonly being from fifty to one hundred and fifty feet high, and from a few hundred yards to nearly a mile in length. After the ice left, the sea rose and flooded the low areas, with the result that the tops of many of the drumlins now protrude from the water to form the elliptically shaped islands which now dot Boston Harbor.

An excellent view looking north over the Harbor can be obtained from the summit of Weymouth Great Hill. From the top of this drumlin a be-wildering display of islands, estuaries, promontories, beaches, and straits unfolds to the north, and we can see coastal land in various stages of being made over by wave action. In the foreground toward the east is Grape Island with a small cliff notched on its seaward edge. Close by is Bumkin Island, with its beautiful smooth shape looming out of the water, looking like some gargantuan sea turtle afloat, basking in the sun. Ped-docks Island a little farther off toward the north is composed of a num-ber of drumlins joined together by sand beaches; the cliffed edges of two of the drumlins show the source of much of the material that makes up the beaches. Farther off to the east, the rounded tops of several drum-lins show up against the horizon; they now form the high rounded parts of the peninsula of Nantasket Beach and Hull. This peninsula has very roughly the shape of the number 7; with its base attached to the main-land, it extends northward for about four miles and then westward for about a mile and a half more. Its principal feature is a fine sand beach which connects the drumlin hills and the mainland.

From the road along Nantasket Beach it is easy to see that the east-ward sides of the drumlins have been chopped off by the action of storm waves. The material which formerly made up the missing parts has gone in great measure into building the beach. Such a sand bar which connects a former island to the mainland is called a "tombolo." In general, the production of any tombolo implies first the formation of an island, then the joining of the island to the mainland by a sand or gravel beach, built by waves and currents. On the north side of Boston Harbor, a counterpart of Nantasket Beach exists in Lynn Beach, which has tied the rocky islands of Big and Little Nahant to the mainland.

The sea still has a job to do in the area of Boston Harbor before results of its activity dominate the landscape and evidence of the drowning is altogether destroyed. Just as in Maine there are the estuaries, such as the drowned seaward ends of the Neponset, Charles, and Mystic Rivers. The coastal scenery here, however, far more than in Maine, emphasizes the work of the sea. Many beaches and sand bars and the wave-cut seaward margins of the drumlins are very important aspects of the landscape.

Traveling along the coast south from Boston, we come to Cape Cod, a peninsula surrounded and largely dominated by the sea. Jutting out miles from the mainland into the Atlantic, this land is completely exposed to the ocean's relentless activities. Made up originally of debris left by the glacier, the Cape is easily eroded; nowhere can we find the smallest ledge of bedrock. Wherever we visit along its shores there is evidence of the ocean's work. The partially destroyed headlands and the precipitous

PROVINCETOWN

LONG POINT

PLYMOUTH

NAUSET LIGHT

MONOMOY PT.

30 MILES

CAPE COD

cliffs which front the sea reveal the destructive power of waves, and the bars, beaches, and sand spits tell of the ocean's constructive power.

On the Atlantic side of the Cape, where the easternmost bulge of land faces the ocean waves, the sea has been able to do its job of destruction without interruption. Indeed, should we visit Nauset Beach on a gray day of rain and storm we may greatly fear for the permanence of the land and wonder how it has been able to withstand the ravages of the sea as long as it has. In the winter when a northeast gale is blowing and flying spray whips across the beach in horizontal streaks, the coast shakes to the impact of the breakers as they crash on the shore. The roaring sound of grinding sand and pebbles as the waves break and return forms a deep background to the high shrill whistle of the wind. The time to fear for the land is when such gale winds and waves are accompanied by a high tide. Then it is that the cliffs facing the ocean are undercut, landslides occur, and more material is lost from the land and made available for redistribution by the sea.

The coast at Nauset Beach Lighthouse consists largely of ocean-produced features, both erosional and depositional. A line of cliffs faces the sea where the wave action has been most intense. The material eroded from the cliffs has gone into the making of the beaches which extend for miles. These beaches in turn are in the process of being washed northward to form the hook of sand and gravel at Provincetown, and southward to form the Nauset Beaches and Monomoy Point. The dividing line which separates the northward from the southward drifting sand is located somewhere just north of Nauset Beach Lighthouse, roughly where the curve of the Cape projects farthest out into the ocean.

South of this point the drowned nature of the coast is still obvious in the inlets at Eastham and the large expanse of Pleasant Bay. Both of these estuaries have by this time been almost cut off from the sea by the growth of sand beaches. The southward growth of these beaches in the form of sand spits, as well as at Monomoy Point still further south, is very obvious on a map, which shows their northern extremities attached to the mainland while their southern ends terminate as pointed fingers in the water.

Most of the transportation of sand and gravel along the shore is accomplished at times of large waves and strong winds. Whenever waves hit the shore obliquely a slow drift of water in the general direction of the wave motion is set up, which carries the finer material in suspension. Furthermore, the larger particles on the beach itself are subject to a slow and jerky motion along the coast whenever waves hit at an angle. The moving water of these breaking waves carries individual sand and gravel particles up the beach as well as slightly along it, and then pulls them straight back down the slope with the undertow. In this way as each of these peripatetic particles is moved up and then down the beach slope, or away and towards the water, it also moves in a direction parallel to the coast a short but measurable amount. Progress measured in terms of thousands of feet per day has been reported for material moved in such a fashion.

The coast extending northward from Nauset Beach Lighthouse to the tip of the Cape is entirely sea-formed and sea-controlled. A continuous beach which hugs the shore ends at the westernmost extremity of Cape Cod at Race Point Lighthouse near Provincetown. For the last eight miles it is a large sand spit, attached at one end to the mainland and built northwestward by the action of waves and currents. More recently currents in Cape Cod Bay have moved some of this material again and used it in building Long Point, a hook-shaped deposit of sand which partially encloses Provincetown Harbor and which points eastward or in exactly the opposite direction from the principal hook. Thus, we see that there has been a two-stage development of the present landform, the initial production of the major spit and then the formation of the parasitic Long Point.

The Pilgrim Monument in Provincetown stands a little way outside and above the center of this picturesque old town. Aside from its historic interest it is well worth a visit, for from the top we can get a splendid view of the end of Cape Cod and can see the sweep of the land toward Cape Race, as well as the reverse hook of Long Point. From the monument, on a clear day, the whole curve of the inside of Cape Cod can be traced southward and then westward, as far as the higher parts of the

SAND DUNES NEAR BARNSTABLE, MASSACHUSETTS.

Courtesy Massachusetts Department of Commerce

mainland near Plymouth to the west of us over twenty miles away.

To the north we can see a very fine collection of sand dunes. Blown by the winds, the sands are often piled many feet above the reach of storm waves. They cover an area a mile or so wide on this seaward part of the Cape and extend for miles along the shore. The very irregular pattern of a typical dune field is obvious. Some small trees stand in the hollows, and grass sketchily covers the dunes themselves. Such a vegetation cover may help to anchor the dunes in position for a while, but only temporarily, as the constant supply of new sand quickly buries and destroys the grass and trees.

Dunes, with their changing shapes and constant drifting, are among the most mobile of all landscape features. For their formation the winds which shape them need only loose sand. In desert areas, the winds are continually at work rearranging the sands that are not held down by vegetation. Along ocean and lake shores dunes are found wherever waves and currents keep washing up fresh and plentiful supplies of sand.

Cape Cod offers many vantage points from which sand dunes can be seen forming an important part of the landscape. The main road eastward, just past the Cape Cod Canal at Sagamore, traverses for a distance the central higher parts of the Cape. From this road a whole series of gleaming white dunes along the shore at Sandwich are visible to the north. Beyond them lie the beach and Cape Cod Bay, its wrinkled surface reflecting the sun in a myriad of shining dots. To the south there is a typical Cape scene of rolling knobby hills, many small lakes, and scrubby pines growing everywhere in the sandy soil.

Leaving Cape Cod and moving farther south down the Atlantic Coast, we find that the geological pattern made by Pleasant Bay and its ocean barrier of Nauset Beach is repeated in many other places. The shores of

Rhode Island, Connecticut, and Long Island are all characterized by the presence of estuaries, spits, and sand bars. The depositional features are most extensive wherever soft glacial material is present, and less so where

SAND SPITS AND BARS ON THE MASSACHUSETTS COAST. Martha's Vineyard is at the left, and Chappaquiddick Island at the right. *Photo by Laurence Lowry*

the waves and currents attack land composed of hard bedrock. Long Island has practically no bedrock, and its southern shore is remarkable for the spits and bars which extend almost all the way from Rockaway to Montauk Point.

South of New York, the coast of New Jersey is practically one continuous sand beach, backed for long stretches by low wave-cut cliffs, and at other places by marshland or lagoons and drowned river valleys.

If we go to the lookout point at Atlantic Highlands, which overlooks the northeastern part of the New Jersey shore, we can obtain an excellent view of Sandy Hook, another example of a large sand spit. From this vantage point, hundreds of feet above the level of the sea, we can note how the Hook attached to the mainland at our feet makes a long reach northward into the sea, stretching almost halfway to Brooklyn, which can be seen twelve miles away across Lower New York Bay. To the northwest lies the large mass of Staten Island, which seems to merge with the New Jersey mainland, and between Staten Island and Brooklyn we can discern the Narrows, the mile-wide entrance to New York Harbor. To the east stretches the open Atlantic.

Sandy Hook was built up by the slow but continuous addition of sand brought by waves and currents from farther south in New Jersey where the ocean has been tearing away the land. At each stage of its growth the spit ended in a hook curving toward the west, just as it does at present. The ends of two earlier and smaller editions of Sandy Hook can be noted on the west side of the present land form where the peninsula is visibly wider. A small parasitic spit similar in origin to Long Point near Provincetown is easily seen from our lookout point. It projects southward and lies on the protected west side of the Hook.

In most cases sand spits are hook-shaped because the waves and currents on reaching the end of a spit tend to carry their load of sand and gravel around in the direction of the major wave and current action.

The northern New Jersey shore is composed of weak layers of mud, sand, and gravel, and as the sea wears it away small cliffs only a few feet high are produced. A great deal of trouble has been taken to protect the coast here from further depredations of the sea; concrete walls have been built and stretches of the shore have been plastered with large blocks of stone. Breakwaters or groins have also been built straight out from the shore to curb the drift of sand along the beach. In spite of all this effort, however, every few years there is a storm powerful enough to toss about the large blocks of stone and to undermine the concrete walls, and a few more inches or perhaps feet of land are surrendered to the sea.

There are still obvious estuaries along the New Jersey coast, although enough time has elapsed since this area was drowned for a number of the smaller ones to have been obliterated. Offshore bars are especially noticeable near the central part of the state. Here the waves have scoured up the shallow sea floor to build a string of barrier beaches a mile or more from the mainland, such as that at Atlantic City. A lagoon of quiet water has thus been formed between the wave-built bar and the mainland. In time the beaches will be pushed shoreward as storm waves toss sand and gravel into the lagoon behind, and at the same time stream deposits will fill in the estuaries and lagoons from the land side. The ultimate fate of this coast therefore will be the eventual obliteration of the lagoons and estuaries. There will be a low intermittent wave-cut cliff, fronted by a sand beach extending from one end of the state to the other; all the estuaries and the lagoons will have been filled. The prevailing scenery then will be wholly a product of the sea.

The same general relationship of offshore bars somewhat masking a drowned coastline is apparent along many stretches of the Gulf and South Atlantic coasts. The coast of North Carolina affords a classic example of scenery due to drowning of land followed by the production of offshore bars. Albemarle and Pamlico Sounds are sheltered behind the barrier beaches culminating in Cape Hatteras. In places these beaches lie twenty-five to thirty miles away from the mainland. They are crowning examples of the depositional work of the sea.

Behind these offshore bars, all along the Atlantic coast from New Jersey to Florida, there are many estuaries. We can note at once that they look quite different from those we saw in Maine. Chesapeake Bay, for instance, has an extremely intricate shoreline, thoroughly ramified by almost countless inlets. Here, along the edge of the coastal plain, the ocean has flooded a river system whose pattern before drowning was dendritic, that is, it resembled very closely the shape of a tree, with main trunk, branches, and twigs. The principal channel of the drowned Susquehanna River follows the central part of Chesapeake Bay and the major tributaries are located where the principal offshoots of the Bay occur.

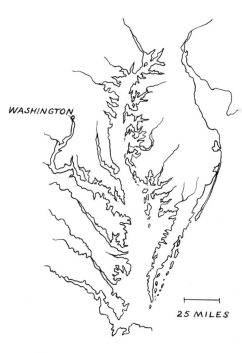

WASHINGTON

25 MILES

CHESAPEAKE BAY

We are conscious of water all around us in this region. From any of the small hills inland we have a view over fertile country, tree-clad and green, and always there is the sparkle of water to catch the eye and mark the course of some tidal inlet. Along the eastern shore the land is flat, a region of level fields and pinewoods. Here there are salt marshes and tidal creeks; everywhere the land has been cut into jigsaw pieces by numberless waterways.

When this part of the coast was drowned, the sea meticulously flooded each tributary and brook of the existing river system to form a new, incredibly complex shoreline. As drawn on a map this shoreline has a fine feathery appearance, due to the tremendous number of small streams which the sea invaded. Along this new coast the sea now touches land which at one time was miles from the edge of the ocean.

IT IS strange that the explanation for the shape of the contact between land and sea in so many coastal areas does not lie in the work of the sea at all, but in the work of some other agent totally unrelated to forces of the ocean. The topography of the original countryside produced by a river or a river system dictates the shape of the coastline when the region is drowned by the sea. Similarly, the shoreline around coral or volcanic islands which have built themselves up out of the water has been shaped by agents other than waves or currents. The sea has merely outlined these topographic shapes produced by other forces.

And so it is with some of the shore forms in the Gulf Coast region. Here we find various oddly shaped bulges on the shoreline where meandering rivers empty into the Gulf of Mexico. These are deltas which have been built of material carried by these rivers. Of all the deltas along this coast, the one built by the Mississippi River is the most impressive.

Currently, the Mississippi River is dumping over one million tons of material a day into the sea. Throughout the centuries it has built a tremendous mound of this material, of which the volume above water is almost an insignificant part in comparison with that which is beneath. Above sea level the delta extends over one hundred miles out into the Gulf, the underwater portion extending of course many miles farther. Furthermore, exploratory oil wells have gone through thousands of feet of deltaic sediment on the coast of Louisiana. For a long time now the land here has apparently been sinking just about as fast as the Mississippi River has been piling up new material.

THE rugged granite along the jagged coastline of Maine, the peculiar oval islands, sand promontories, and beaches of the Boston area, the extensive offshore bars, beaches, and lagoons of the South Atlantic and Gulf

Coasts—these are the varied features of our rambling eastern and southern coastline. For an utterly different kind of shoreline, one which is produced primarily by wave erosion, we must turn to the Pacific Coast.

A casual glance at a map tells us that in general the western edge of our country is fairly straight, with contact between sea and land running north and south. Scenery along the Pacific Ocean consists in large part of wave-cut cliffs, raised terraces, small rocky islands near the shore, and short lengths of beaches between rugged headlands. The only estuaries on the West Coast which compare in size with the deep tidal inlets of the East are those which form Puget Sound and San Francisco Bay.

On a misty July morning, as the fog begins to lift, Cape Sebastian on the coast of southern Oregon affords a superlative vantage point from which to view marine destruction at work. Taking a short trail, sheltered for the most part from the ocean wind, we come to the end of this headland, well out in the Pacific. From this point hundreds of feet above the sea, we get a view of the coast toward the south. The land drops steeply, in places almost precipitously, to the water's edge. The slopes are covered with tawny grass, with here and there some scattered conifers and myrtles. In the nearer distance some rocky islets dot the sea, their jagged shapes outlined against shifting patches of fog. Farther toward the south, through a break in the lifting fog and near the next headland, an unmistakable sea arch is visible. Here whitecaps can be seen glinting through a large hole cut by the waves into an isolated rock located well offshore. The whole composes a gray, stupendous scene of destruction, breaking waves, and the remnants of a vanished land jutting out of the water. The islets, in process of final obliteration, and the receding precipitous cliffs of the land itself, have but a short life remaining.

An inspection of the rocks of the headland of Cape Sebastian shows that they are composed of a well-cemented sandstone, whereas to the side and inland, the sandstone is more crumbly and is more easily eroded because of the greater number of shale layers present. The attack of the ocean on such a land is determined and unceasing. The more resistant parts will remain intact a little longer than the softer material and will form steep cliffs, headlands, and rocky islets, known as sea stacks or, more simply, stacks.

SEA STACKS ON THE OREGON COAST. Such rocky islands, once part of the mainland, have been isolated by the erosion of the sea.

Courtesy Oregon State Highway Commission

We know that the edge of the land once extended beyond the furthest of these sea stacks, and that now they merely indicate the minimum extent formerly reached by this part of the continent. The islands of Maine also show the former extent reached by the mainland, but the origins of the islands of these two coasts are utterly different. Those in Maine, as we have seen, are hilltops first molded by stream erosion and then isolated from each other by the rising sea. The stacks, on the other hand, are more resistant parts of the land which have withstood the forces of wave erosion a little longer than the material which at one time surrounded them. Compared with the islands made by drowning, sea stacks are generally smaller, nearer the shore, and obviously associated with a receding wave-cut cliff.

From our vantage point, high up above the sea, we can see the procession of waves as they move in toward the shore forming an ever-changing pattern. Out to sea the wave fronts are straight, but as they near this curving shore they bend, as though each wave were attempting to break head on against the entire coastline with a single impact. The waves are forced to bend in this way as they maintain their speed into the coves but slow up in front of the promontories where the water shoals. They do not break with equal force against every point along the coast. Every wave has a certain amount of energy of water motion along each foot of its length. When waves are bunched together opposite a promontory this energy is concentrated and a far greater share of it is expended on the rugged headlands and stacks than is ever spent in the coves, where the wave is stretched so as to cover the whole length of the shore.

The process of wave-bending or refraction and the resulting concentration of energy on the promontories explains why indentations on wave-cut coasts are seldom more than a fraction of a mile deep; they never reach the size of estuaries on a drowned coast.

Here and there along the Pacific Coast the traveler's attention is drawn to cliffs whose tops are absolutely flat, as though carefully leveled by a bulldozer. These are actually wave-cut terraces, a part of the land flattened by the ocean before a drop in the sea level occurred. Such a feature is displayed to perfection in the Coos Bay area of Oregon. At Shore Acres State Park, for example, the visitor can stand about sixty feet above sea level on a beautiful flat terrace between four and five hundred yards wide. At the inner margin of the flat area, the land rises abruptly and steeply. This sudden rise marks the former edge of the land where waves once broke. The waves now break sixty feet lower, and are in the process of cutting a new terrace, and a new cliff facing the sea is taking shape.

In this same general area of Oregon, remnants of wave-cut terraces have been found at even higher elevations. A succession of such features can be noted to an elevation of over 600 feet on Seven Devils Hill just to the south of Coos Bay, and on a hillside farther inland there is a terrace deposit at a height of 1,500 feet. These old beach deposits on the raised terraces contain a variety of minerals. At one place gold was found in marine sands about 200 feet above present sea level, at the base of an old sea cliff, concentrated there by the sorting action of waves and currents.

To the south of Cape Sebastian driving on the coast road we see a continuous panorama of changing scenes due primarily to marine destruction —cliffs, with usually a narrow beach of sand or gravel at their foot, sea stacks, and raised terraces. Near Brookings, Oregon, a little hill can be noted rising from a raised terrace. This can be interpreted as a stack on a former bit of the sea floor. Though not too common, such features can be found at a number of places on the Pacific Coast.

Between the headlands there is generally a greater accumulation of beach material, enough at times to form quite an impressive deposit. At

PALOS VERDES, CALIFORNIA. The series of terraces were cut by ocean waves when the land was lower. A remnant of a recent raised terrace backed by a wave-cut cliff is especially obvious at the lower left. Now the sea has produced a new wave-cut cliff and is widening a new and lower terrace. Note also the refraction of the waves as they enter the bays. *Spence Air Photos, Los Angeles*

rarer intervals estuaries, many of which are now nearly filled by stream deposits, remind one strikingly of the drowned valleys of the East Coast. Bars may frequently extend across former estuaries, thus producing stretches of land-locked waters. Classic features of this type are located near Orick, on the northern California coast; here bars with beautiful slight curves cut off Stone Lagoon and Big Lagoon from the open sea.

On both the East and West Coasts of the United States there is abundant evidence in the form of drowned river systems or raised wave-produced features, to demonstrate the rising or falling of sea level with respect to the land. Such a change can be produced in either of two ways. A general worldwide change of sea level may occur or a local shifting of rocks in the crust may take place whereby a part of a continent is raised or lowered with respect to adjacent parts. It may be impossible to determine the cause of a specific shift in sea level. For instance, a river is drowned whenever the ocean rises or the land sinks, and either process will produce identical results.

In the next chapter we will see that the large-scale development of glaciers on the land and their subsequent melting has had very marked effects on sea level. The most recent drowning of the coastal areas of the world is due to the return of a great deal of water to the ocean, water which had been locked up on land in the solid state.

ROCKY PASTURE, CORNISH, MAINE *Courtesy Maine Development Commission*

CHAPTER 3

NEW England is a rocky land. From the seacoast to the mountaintops the traveler is continuously reminded of this fundamental scenic theme. Gray rock ledges protrude through an uneven coating of stony soil. Marking lands once cleared so laboriously by the early settlers, stone walls of weathered, lichen-covered boulders crisscross the countryside in many directions. They can often be followed for miles along the back roads and

The Legacy of the Glaciers

frequently disappear into dense woodland patches, lost to sight in the vegetation that has grown up around them. In the lowest places of this rocky region a multitude of various sized lakes and swamps are found; the highest parts are in such pleasant mountainous tracts as the White Mountains, the Green Mountains, and the Berkshire Hills.

The landscape of New England has counterparts in many other places in the world. Before one can fully understand it, however, like an artist one has to learn to *see* the land to recognize its underlying nature and peculiarities.

Stopping now and then in our travels for a closer look at the obvious features, we come upon evidence which has a direct bearing on New England's geologic past. The evidence is everywhere, from the Atlantic shore to the Presidential Range of New Hampshire, where the White Mountains reach their highest elevation in Mount Washington, and its attendant peaks, Adams, Clay, Madison, Monroe, and Jefferson, all of which stretch beyond the timber line in boulder-strewn crests.

As we drive across the country we notice a number of low gravelly ridges and hills, now covered by vegetation. If we stop for a moment where a road has been recently cut through one of them we can see what their underlying material is like. Surprisingly, it will often be found to consist of sand and gravel which strongly reminds us of the Maine coast. There are the same rounded and obviously water-worn pebbles, and in places a sorting of the material can be noted which resembles the sorting of sand and gravel on the beaches.

These observations, as homely as they are, had scientists baffled for many years. How were these piles of sand and gravel formed? Many are much too far away from any modern stream to have been dumped at a time of high water.

Until the middle of the last century, it was believed that they were the result of a general flooding of the land. This flood was not caused by a rising sea or the swelling of rivers, but by a wall of torrential waters rushing and tearing over the land, inundating whole tracts of the countryside. Where the water came from was of course a problem. One explanation was that it had been stored for a long time in large caverns under the sur-

GRAVEL PIT IN A LARGE ESKER NEAR OSSIPEE, NEW HAMPSHIRE. Note the rounded water-worn pebbles and the absence of mud. The top of the esker ridge is shown and the start of the slope down on either side. *Photo by John Shimer*

face of the earth and that it was blown out by some cataclysmic event. Attributing the sand and gravel piles to such a flood seemed doubly logical because many folklores include a Great Flood in their setting. Such a flood could also explain the presence of most of the numerous boulders scattered over the countryside. But some of these are of fantastic size and weigh many tons. One of the largest, the Madison Boulder, which is located a few miles south of Conway, New Hampshire, is a block of granite which weighs in the neighborhood of 4,662 tons and measures 83 x 37 x 23 feet. A twenty-room house would fit into these dimensions very nicely and leave plenty of room for a large garage besides. The parent ledge from which this massive block came has been located. It is roughly two miles north of its present position. It seemed inconceivable that such a boulder could be transported over such a distance, no matter how powerful the rushing flood. Some scientists therefore suggested that icebergs could have rafted the large rocks over the flooded country, dropping them finally where they are found today.

Leaving this problem unsolved for the moment, let us return to the road cuts and make further observations. Some cuts will expose, instead of the sand and gravel, a different kind of mixture, one in which fine mud is also present and is intimately mixed with the coarser material. This mixture, moreover, obviously cannot be soil which has developed from the weathering and breakup of the underlying bedrock because the pebbles are of stone totally unlike the rock ledge on which they rest.

Nowhere in the world today, either in a stream or along the seashore, can one find such a mixture being deposited by running water; flowing water sorts the material it carries. The carrying capacity of a stream becomes less as its velocity decreases, so its burden is dropped in successive stages, the large boulders first, then the sand, and finally the mud. For the same reason beaches, too, are generally composed of well-sorted layers of material. Thus, for the explanation of a heterogeneous mixture of materials we must look elsewhere than in the power of running water.

In addition to these mysterious soil deposits, there is another odd feature in this stony land which intrigued scientists for many years. On many of the bedrock ledges there are fine parallel scratches, and in some places rather deep grooves. These grooves and scratches are often found crossing any banding or layering inherent in the rock. Moreover, they are associated with the smoothing and rounding of the top surfaces of the bedrock.

The true explanation of these incongruous features in New England came in 1846. If light on the problem had not come indirectly from the Swiss Alps it would perhaps have taken many years longer.

Between 1821 and 1835, two European geologists, J. Venetz and Jean de Charpentier, suggested that the current glaciers in the Alps had at one time been far more extensive and had in fact covered a great deal of Switzerland. To prove their point they first called attention to the well-known fact that the Swiss glaciers as they flow down the valleys carry a large amount of debris either frozen in the ice or riding on its surface. This debris varies in size from enormous boulders to very fine sand and mud. Furthermore, it was noted that as the glacier melts on reaching lower elevations it drops its load into piles consisting of all sizes of debris. Also, wherever the glacier has melted back, the bedrock of the valley is scratched and grooved, very obviously by boulders frozen into the ice that has been moving downhill.

No one doubted that a glacier could be responsible for such piles of debris, large boulders, and striations found in its close proximity. However, Venetz and Charpentier further suggested that such features found many miles away from the fronts of existing glaciers indicate the former extent to which the ice has reached, and that it has left its path strewn with similar evidence of its passage. Other geologists who had not seen the evidence greeted this explanation with a great deal of skepticism.

A Glacier-Borne Boulder, Cohasset, Massachusetts. *Photo by John Shimer*

Louis Agassiz, a young Swiss scientist, who later became one of the best-known geologists of his time, was at first one of the skeptics. But when he saw the evidence in 1836 he became firmly convinced that Alpine glaciers had in fact at one time been far more extensive. He even went a great deal further than this; he postulated that similar features in England and the Eastern section of the United States reveal that parts of these areas also were at one time covered by ice. This truly revolutionary suggestion called for the former spread of glaciers in lands and over areas where there are none at present. So it was no wonder that the general acceptance of such an ice-age concept was very slow in coming. However, it became absolutely inescapable as feature after feature was observed which could only be explained adequately by such a concept. It was in fact primarily due to Louis Agassiz, geologist and zoologist, that the well-nigh incredible idea of a glacier covering New England in former times came to be at first entertained and then at last accepted.

With this idea in mind, the motorist driving through New England will see how neatly the pieces of the puzzle fall into place. Ice can freeze onto all kinds and sizes of material and on melting will drop its load into a pile of unsorted debris, or "till," in which coarse gravel, sand, and mud may be mixed. The boulders, large as well as small, now found littering the countryside or neatly piled up in the walls laid out by the early settlers could have been moved very easily by the ice and then dumped when it melted. Some of them appear so precariously perched that it seems as though the slightest touch would topple them over. The scratches and grooves which are found on so many bedrock surfaces and which run, in general, north to south, are now easily explained. When the ice moved over these ledges any sand or gravel embedded in its lower surface acted as abrasive tools to scrape and scratch the bedrock.

The haphazard dumping of material by the melting mass of ice inevitably produced an irregular surface, leaving hollows which became lakes and swamps; these are very characteristic features of glaciated regions. To be sure, layers of sorted sand and gravel are also common in New England. These are explainable as water-laid deposits, the sand and gravel having been carried and sorted by flowing water from the melting ice. Any mud present was washed into glacial lakes or the sea, leaving primarily the coarser material behind. With such large quantities of melt water it is really somewhat surprising that there are as many deposits of till as there are.

GLACIAL STRIATIONS ON MOUNT KEARSARGE, NEW HAMPSHIRE. They appear as surface gouges and scratches almost parallel to the handle of the hammer.

Photo by John Shimer

The time which has elapsed since the melting of the glacier, about ten thousand years, has been adequate for the agents of rock weathering to partially destroy many of the striations. However, where they have been protected by a layer of soil and but recently uncovered, or where they were deep in the first place, they are perfectly obvious and sometimes apparently as fresh as the day they were made.

Some truly wonderful glacial striations and grooves can be found on the upper ledges of Mount Kearsarge, located about seventy miles south of Mount Washington. The path to the top which starts from the northwestern foot of the mountain at first climbs gently through a forested area liberally sprinkled with boulders. Soon the way steepens and we encounter rock ledges, which continue to the top. It is on these ledges, particularly near the summit, that evidence of glacial erosion is most apparent. One of the most impressive features of the scratches and grooves is the consistency of their orientation. They all run roughly in the direction we are climbing, and when we must scramble up a difficult ledge they often run right up the rocky surface with us. As we near the top the depth of the grooving becomes greater and another feature of the rock ledges becomes more apparent. The northwest side of each one of these ledges is much smoother and more streamlined than the southeast side, which generally has a rough, jagged outline. Now if we visit similar ledges on the other side of the rounded rocky summit we find that the northwest faces are still the streamlined ones. These asymmetrically shaped ledges obviously do not owe their origin to the slope of the land but to some agent which attacked the rock from one direction only. In this case the agent was ice, which came from the northwest and streamed toward the southeast. The glacier rode up over the northwest face of each ledge, smoothing and gouging the rock as it did so; the jagged southeast face resulted from the plucking action of the ice, which froze onto bits of broken or frost-shattered rock and carried them away.

The fact that striated rock is present at the top of this 2,937-foot peak indicates that the ice surface here was at one time well over 3,000 feet above sea level. An awe-inspiring scene certainly must have existed at that time. We can only imagine the cold snowy world which consisted of ice streaming southward with not a bit of solid land in sight. A little later when the flood of ice started to subside and the higher parts of the land began to emerge through the thinning sheet, the view from the top of Mount Kearsarge would have been still more unreal. Islands of dark gray rock would have appeared, jutting out of the white world still existing in the valleys.

The ice mass which overrode the top of Mount Kearsarge also covered for a brief time the highest part of New England, the 6,288-foot summit of Mount Washington. It left till and boulders to mark its presence. The highest glacial grooves are found near the Old Crawford Path at an elevation of 5,700 feet. Any exposed striated ledges which might have been left nearer the summit were broken into angular fragments by the intense post-glacial frost action occurring at this elevation, which also redistributed the till and boulders. Today the top of Mount Washington is heaped with rocks which completely cover the underlying bedrock. The last part of an ascent by foot necessitates a chamoislike ability to step from the top of one rock to another.

Such an ascent might well be made from Pinkham Notch on the east side of the mountain. From here an excellent trail climbs roughly 4,000 feet via Tuckerman's Ravine to the boulder-strewn summit. The way at first follows a mountain stream and the small valley it has cut. This valley, like all proper stream-cut valleys, gets gradually smaller as we climb into

MOUNT WASHINGTON, NEW HAMPSHIRE, showing the cirque of Tuckerman's Ravine in the foreground. *Photo by Laurence Lowry*

its headwater area. However, somewhat less than half way up the mountain it suddenly opens out into an utterly different kind of valley with a more gentle slope and a wide floor. Here in the valley floor is rockrimmed

Hermit Lake, almost hidden in the thick forest cover. We follow the trail past this relatively flat part to a very short steep slope, after which there is again a lessening of the gradient, and we step out onto a large open amphitheater, Tuckerman's Ravine. We have at last come to the head of the valley, but what an unexpected contrast to the valley at the base of the mountain. Here is a cirque with its typically large spoon-shaped hollow scooped out of the mountainside. Very steep sides flatten out gradually to the rounded bottom, which in turn slopes gently downward to the abrupt descent over the Little Headwall and to the secondary lower valley of Hermit Lake.

Another view of this truly amazing hollow can be obtained by following the path as it zigzags up the headwall to the west of us, until we reach the top and can look down at the gouged-out mountainside. The summit of the mountain behind us is still a thousand feet higher, but the going thither is easy in comparison to the steep ascent just made. We are above the timber line, and the view is extensive. In the summer the thick vegetation cover in the valleys and lower flanks of the mountains shows contrasting shades of green, the darker patches of evergreens interspersed with the lighter green deciduous trees. With the coming of autumn and cooler days the whole scene changes and there are some of the most colorful sights imaginable. Winter brings a thick blanket of snow. In the Ravine, located as it is on the east side of the mountain, the snow stays well into the spring months, and in some years patches have persisted into the heat of August.

The fashioning of Tuckerman's Ravine started when during the winter more snow drifted into a shallow stream-cut crease in the mountainside than could melt during the following summer. Year after year the snow falls of the previous winters were changed into ice by the compacting of the buried snow as well as by partial melting followed by refreezing. Such a process is illustrated by any bank of snow which has lasted into spring or early summer and which will be composed of heavy granular particles, the result of this melting and refreezing process. Eventually, the pile of ice on the mountainside became so large that it started to ooze down the hillside under its own weight, following as it did so the line of least resistance, which was the narrow preglacial stream valley. As the upper end of the mass of ice pulled away from the mountainside it left an opening between the ice and the bedrock. Melt water streamed into this fissure by day and froze by night, with the result that the surface of the bedrock was shattered by frost wedging. Finally these broken fragments were then frozen into more snow and ice as it formed in the crack and were removed by the forward motion of the glacier. In this way, by freezing onto shattered rock at the upper end and gouging out the base the glacier literally ate a hole for itself in the side of the mountain, and thus formed this cirque of Tuckerman's Ravine.

We can imagine what this scene must have been like in the past by picturing the hollow filled again with ice. The glacier as it fell over the Little Headwall probably presented a scene of wildly tossing blocks deeply crevassed and very difficult to traverse. Next the ice stream flattened out as it flowed more easily over the bedrock, scouring out as it did so the small basin now filled with water to form Hermit Lake. We can imagine the glacier heading down toward Pinkham Notch but not getting very much farther since at these lower and warmer elevations the rate of melting equalled the rate of supply of new ice coming from above.

Ice is a peculiar substance. Under normal conditions it is brittle and yields to stress by breaking, like glass. However, under pressure and slow deformation it can be made to flow in a fashion resembling a thick viscous liquid. This flowage occurs only in the underlying layers, which are subject to pressure from above. The upper layers of a glacier ride on top of the underlying flowing ice and conform to valley shapes by fracturing and by the slipping of one block relative to another. Crevasses in the ice do not usually stay open to a depth of more than about two hundred feet at most. Below this the pressure of the overlying material is adequate to push the sides of any crack together, thus sealing the opening.

The rate of flow of the glacier in Tuckerman's Ravine was undoubtedly slow. The actual rate of travel of a glacier may vary between a fraction of an inch to about ten feet per day, depending on such variables as angle of slope, thickness of the ice, and temperature. Occasionally misconceptions have arisen regarding the speed with which a glacier travels, although not all are as extreme as Mark Twain would have us believe, in his *Innocents Abroad*. In a delightful account of his experiences in the Alps, he tells of one occasion when he and his friend camped overnight on a glacier, with all their baggage, expecting to be transported rapidly to their destination as they slept on the moving river of ice.

Wherever we find valley-type glaciers today, the ice tongues always start well up in the mountains at an elevation where precipitation of snow is greater than the rate of melting. Under such conditions a glacier is inevitable and invariably will cut for itself a hollow in the side of the mountain, which, when the ice eventually disappears, will reveal itself as a cirque, with a hollowed-out floor in which a small lake, or tarn, may form. In the case of Tuckerman's Ravine there does not happen to be a lake in the cirque floor itself but as we have seen the glacier did dig a small hollow farther down which now holds Hermit Lake.

Tuckerman's Ravine is only one of a number of cirques which head in the mass of the Presidential Range. Huntington Ravine lies on the east

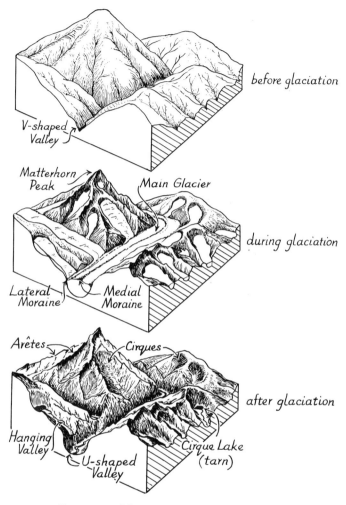

before glaciation

V-shaped Valley

Matterhorn Peak
Main Glacier

during glaciation

Lateral Moraine
Medial Moraine

Arêtes
Cirques

after glaciation

Hanging Valley
U-shaped Valley
Cirque Lake (tarn)

EROSION OF MOUNTAINS BY VALLEY GLACIERS

side of Mount Washington, just to the north of Tuckerman's Ravine, and King Ravine is on the north side of Mount Adams. The Great Gulf separating Mount Washington from the mass of Jefferson and Adams, which begins as a large cirque on the flank of Mount Clay, was shaped by ice flowing down from a number of places of accumulation on the neighboring mountainsides.

If these valley glaciers, both before and after the continental sheet covered the whole area, had been given more time to cut into the mountain mass, a far more jagged type of scenery would have been produced, a landscape resembling the Alps much more closely. The cirques would have grown wider and deeper, and eaten more and more into the mountain mass itself, until eventually none of the original rounded dome of Mount Washington, for example, would have been left, but only a jagged spire from which one could look directly down into cirques on all sides. Had this happened Mount Washington today would resemble the Matterhorn Peak of the Alps. Furthermore, the ridges between the cirques would have been reduced in width as the cirques became wider until only a jagged divide would have been left. Such a feature is called an arête, a word from the French Alps which is now used for similar features elsewhere in the world.

A landscape of cirques, Matterhorn peaks, and arêtes with well-developed glaciers still present in the cirques from which they stream down valleys, is found at the present time in the Alps and the Alaska Range, among other places. Landscapes with merely patches of glacial ice still in some of the cirques are found in the Rocky Mountains in Montana, the Grand Tetons of Wyoming, and in the higher parts of the Sierra Nevada in California. This jagged, fretted type of upland topography is very characteristic and readily recognized as the work of valley glaciers. When the glacial attack has been less long or extreme a rounded upland surface with cirque bites taken out of it is the result. The Mount Washington area, the Uinta Mountains of Utah, and the Bighorn Mountains of Wyoming exhibit this type of topography in places. The cirque is probably the most distinctive and readily identified of all glacial features to be found in mountains. It is very common and its characteristic shape can be recognized from a distance of many miles. In fact it is almost impossible to drive or climb anywhere in a glaciated mountain area without seeing them.

There is only one basic requirement for the initiation and growth of a glacier. It is simply that on the average, more snow must fall during a year than melts. If this situation persists long enough a glacier is the inevitable result. This requirement can be met by climatic changes which will either increase the amount of snow precipitation or decrease the average yearly temperature, or do both. A generally acceptable explanation for such requisite changes in climate is still very much open to speculation at the present time. Suggestions range all the way from shifting poles, to changes in the earth's atmosphere, to changes in the amount of heat given off by the sun.

In general, the higher the elevation or latitude, the colder it gets and the more likely glaciers will tend to form if the required precipitation of snow occurs. The lowest elevation at which there is perennial snow, near the equator, is over two and a half miles, and only the mountains over this height will possess glaciers. The snowline or elevation at which there is perennial snow gets lower towards the poles until it is at sea level near latitude 60 north or 60 south. Where precipitation is heavy the snowline is lowered for a given latitude. For example, in the St. Elias Range on the Alaskan-Yukon border it is about 3,000 feet in elevation on the western side where precipitation is heavy and it is 8,000 feet on the drier eastern side. There is a large tract of land in Central Alaska, in the Yukon Basin between the Brooks Range to the north and the Alaska Range to the south, which has never been glaciated, even in the coldest times of the Ice Age.

Both of these ranges were ice covered and the Alaska Range still possesses some very nice valley glaciers, but the snowfall was inadequate for the lowland area to be covered by ice, either growing locally or from the mountains to the north or south. It has been estimated that the lowering of the average yearly temperature need not be great to initiate a new Ice Age similar to those of the past; less than ten degrees would probably be more than adequate.

While we have in mind what a slight climatic change it might take to bring a fresh mantle of ice creeping down over New England, let us take a look at another of the particular features of this glaciated region. As the motorist drives through the Presidential Range of New Hampshire, he will undoubtedly be struck by the interesting local use of the term *notch*. Here it means a pass through the mountains, as at Franconia, Pinkham, and Crawford Notches.

In going through any of these passes a divide or high point is always crossed although it may not be very obvious because the climb from either direction can be quite gentle. The notches are relatively straight and in marked contrast to the generally tortuous paths of stream-cut valleys. The truly unique feature of these notches, however, is their U-shaped cross profile. This is especially well developed in Crawford Notch. It may not be immediately discernible to the average motorist driving along the road, for the valley is thickly clothed with trees, both deciduous and evergreen. But if one takes a trail off to the side and climbs a little way, particularly in late fall or winter when the trees are bare and the rocky slopes are visible, it will be clearly seen that it is indeed a U-shaped valley. The steep upper slopes very gradually become less steep until the rounded valley bottom is reached. Such a valley shape is an unequivocal clue to the former presence of a glacier. It can be produced only by a river of ice flowing through a

CRAWFORD NOTCH, NEW HAMPSHIRE, from the summit of Mount Willard. A U-shaped glacial trough. *Courtesy New Hampshire State Planning and Development Commission*

previous stream-cut valley, riding high as it does so on the valley sides, where it will grind and gouge away at the valley walls as well as the bedrock at the bottom, thus rounding out the U shape. For the same reason the transverse profile across a cirque is similarly shaped.

The sequence of events which produced today's scenery around the White Mountains was probably somewhat as follows. Small glaciers were formed and grew on the upper parts of the Presidential Range while the main continental mass of ice was still miles to the north in Canada. Then the great sheet of ice began to pour into New Hampshire, with white fingers probing the valleys in front of the main mass, the icy flood rising higher and higher. Coalescing with the locally formed valley glaciers, which were streaming down from the high parts of the Presidential Range, it eventually overrode everything, until there was a sheet of ice moving southward as far as the eye could see. It flowed over the top of Mount Kearsarge, over Massachusetts, Connecticut, and Rhode Island and reached as far as the northern side of Long Island, where the greatly attenuated front of the ice mass stranded and melted away into streams of water flooding southward into the ocean.

In these lower and warmer latitudes the rate of melting just equalled the rate of advance. The ice front at such a time would appear stationary, but would actually be in a state of dynamic equilibrium where the motion forward was equalled by the melting backward. Under such conditions the load of debris which the ice carried would all be dumped at one place to form a large pile of till in the shape of an irregular, hummocky ridge, or terminal moraine. Such a pile would keep growing as long as ice was pushed forward to take the place of that which melted. The melt water from the glacier would only have one direction to go, away from the ice front, and as it streamed away it would carry a great deal of sand and gravel which it would spread out as an apron-like deposit, or outwash plain. The hilly northern side of Long Island which extends from Brooklyn to Montauk and Orient Points marks the southernmost limit in this area reached by the ice sheet. The land to the south of this moraine was never covered by ice. The flat plain which slopes very gently from the ridge area southward to the sea, is the outwash plain. It is composed primarily of layers of sand and gravel, the mud probably having been carried still further southward to lie beneath the present sea level.

The difference in topography between the two sides of Long Island is dramatic. In contrast with the flat outwash plain, the terminal moraine consists of many small irregular hills, up to a few hundred feet high with low areas between them, which are often filled by ponds or swamps. At the western end of the island near New York City there is only one morainal ridge, but toward the eastern end this divides into two, marking slightly different points reached by two advances of the ice. The dual nature of the moraine is neatly emphasized by the drowned shoreline. Orient Point to the north and Montauk Point to the south mark the drowned extremities of the two moraines.

Among the low morainal hills lie the swamps and ponds. Many of the ponds are in "kettle holes." These were formed when large blocks of ice buried in the till melted, causing the subsequent collapse of the surface into roughly circular depressions. Such knob-and-kettle topography is typical of the many moraines found in New England; an especially fine example is the northern hilly part of Cape Cod.

If we go to Coney Island today and in imagination bring the ice back to the point of its maximum advance, we would see to the north of us a high ridge of ice, covered with a great deal of dark-colored debris. Manhattan and the Bronx and all the country farther north in Westchester and Connecticut would be buried under ice, and no one would even suspect the existence of Long Island Sound. To the south the shoreline would be far off in the distance, and we would be standing on a sand and gravel plain flooded by a tremendous network of streams, fed by the melting ice. During the relatively warm days of summer each channel would be overflowing and quantities of mud, sand, and gravel would be in transit southward. Much of this material would eventually reach the sea, but some would be left stranded on this ever-growing outwash plain. The beautiful sand beach which lines Coney Island today would be missing. Its development would have to wait for the return of the sea up the outwash slope and the piling up of sand by wave and current action. A visit to the growing moraine right at the edge of the ice would reveal a heterogeneous mixture of various-sized boulders and pebbles, mixed with sand and mud. When, now, these few thousand years later, we visit this area we can identify many of the boulders as having been brought from ledges many miles to the north.

DRUMLIN NEAR GLEASONDALE, MASSACHUSETTS. *U.S. Geological Survey*

Walking on the busy streets of New York between the towering buildings, we find it difficult to visualize a glacier covering Manhattan Island and its surrounding regions. Nevertheless, evidence of the glacier can still be found. In Central Park there are many rock ledges which show glacial striations. It may take close inspection to identify them because the bedrock here is mica schist, a foliated, or layered, metamorphic rock. The layering which shows on the surface might be mistaken for glacial scratches. However, looking closer we find surface grooving which intersects the

foliated structure at an angle. The glacial effects are purely surface phenomena, whereas the layering is a basic feature of the rock's structure.

Although the bedrock in Central Park is mica schist, we find here and
there boulders of a moderately fine-grained igneous rock, basalt, which is
black on a freshly broken face but which commonly shows some yellow-
brown weathering stains. These blocks were brought by the glacier across
the Hudson River from the Palisades, and here in Central Park they appear obviously alien.

Deposits of till, whether scattered haphazardly as a thin coating over
the land or piled up in moraines, create a hummocky type of topography.
However, where a glacier has overridden a previous deposit of till the distinctive drumlin has been formed and the typical oval shape of such hills
as those forming the islands in Boston Harbor are produced. They are
slightly less steep on the southern side than on the northern end. This
slight difference of steepness is due to the dragging action of the ice as it
pulls away southward from a pile of till it is shaping, in contrast to the
plowing and steepening action on the northern end which faces the on-
coming ice.

When climatic conditions changed and less and less snow survived the
summer in Canada, the motion of the icy blanket over New England slowed
up and eventually ground to a stop. Then for many years the ice melted
and water flowed off the land. As soon as the summit of Mount Washington was freed of the ice sheet the valley glaciers undoubtedly started to
function again and continued until quite recently. New England at that
time was a cold, wet place, and the sound of running water was everywhere. In the beginning only the tops of the hills showed through, but as
time passed, more and more land reappeared, until eventually only string-
ers and patches of stagnating ice were left in valleys and other low places.

It was at this stage that the sorted sand and gravel deposits, which are now so widespread over the land, were formed.

Driving along a road in New England we occasionally come across elongated ridges which at first sight could be taken for old railroad embankments. There are the same steep sides, and narrow top, but the ridges are apt to wind about in a rather aimless fashion, with sinuous curves quite unlike a railroad. Such ridges are called "eskers." Varying in height from 20 to 100 feet, eskers may extend for a number of miles and frequently run through swampy areas. They are the remains of deposits laid down by streams flowing on bedrock in cracks in the stagnating glacier or even under the ice. In either case the sides of the stream beds were formed of ice. The running water laid down deposits between the walls of ice; when the walls disappeared upon melting, the material deposited by the streams was left standing as ridges. A remarkable group of long eskers is located in the low country to the east of the Penobscot River in the vicinity of Bangor, Maine. They extend for many miles in a north-south direction and are in places over 100 feet high above the general level of the country. They are most handsome specimens of eskers, but it is a little difficult to see them in their entirety especially when shrubs and trees cloak and confuse their shapes.

Elsewhere in New England steep-sided conical hills, somewhat smaller than drumlins, composed of sorted and roughly layered sands and gravels might be found. Such features, called "kames," were formed either by streams flowing off the surface of the ice and building up piles of debris right against the ice wall or by the filling of a hole in the ice by sand and gravel carried by flowing water. A kame or an esker makes an excellent source of sand and gravel; both are frequently sought for commercial purposes.

New England is more or less an entity when it comes to glacial features. The emphasis may change from section to section, but the general appearance of the land is consistently one of gray igneous and metamorphic rocks protruding through a stony and boulder-ridden land.

The preglacial appearance of this part of North America was undoubtedly similar to what it is today in its major outline. The mountains and valleys were where they now stand. The changes brought about by the ice involved only surface features. Nevertheless, they must have altered out of all recognition the former appearance of the land. Lakes and swamps were produced in a terrain essentially devoid of them before. The deep preglacial residual soil was ploughed up and redistributed. A great deal of material was brought in from Canada and is now spread out over northern New England and much of the soil from southern New England was carried across Long Island Sound and now forms the moraines on Long Island as well as the outwash material further to the south. Valley shapes were modified into rounded U's, and the profiles of some of the higher mountains were changed from their previous smooth forms by the hollowing of cirques. Glacial deposits in the form of erratic boulders, moraines, drumlins, kames, and eskers now dot the land where such shapes had been entirely absent before. And everywhere the bedrock shows a scratched and grooved surface.

This land in a sense is haunted by ice, and it does not take much imagination to picture it covered again with a thick slowly moving white blanket relentlessly grinding down over the countryside from the north. Weathering and stream erosion must have begun to modify the land immediately on the departure of the ice, and the fact that there is still such an abundance of very easily removed material is indicative of what a short time ago the glaciers left.

This discussion has taken into consideration only one major advance of the ice from Canada. There is evidence, however, that there have been four major worldwide advances followed by times of warmer weather, when the climate was even more temperate than at present. In New England till from two different ages has been distinguished; in the middle part of the continent deposits of four different ages have been discovered, sometimes lying one on top of the other. The glacial scenery which we see, however, was produced by the last ice advance, which overrode and destroyed much of the evidence of the earlier ones.

GLACIAL scenery in the United States reaches some of its greatest and most dramatic aspects in the mountains of the West. It is here that valley-type glaciers still persist in the higher mountains, and it is here also that the past work of such glaciers is most obvious.

Perhaps the best-known and most impressive relic of the Ice Age to be found anywhere in North America is the Yosemite Valley in California. This dramatically incised valley located on the western slopes of the Sierra Nevada started as a mere stream-cut trench on the slope of a rising mountain mass. The magnificent shapes carved from the mountains here by ice flowing through this valley attest the awesome power of a glacier.

The approach to Yosemite Valley from the west is gentle as we follow the Merced River upstream. From the Central Valley of California as far as El Portal the canyon is narrow and winding, and there is barely room for the road as well as the stream. Here the traveler is forever conscious of the steeply sloping rock walls as they descend right down to the road. The shape of the valley up to this point has a rough V-shaped cross profile. Such a profile is characteristic the world over of youthful stream-cut valleys. A few miles above El Portal a change occurs and everything seems to

YOSEMITE VALLEY, CALIFORNIA, as seen from the Wawona Road Tunnel. At the left
El Capitan rises 3,600 feet above the valley floor, Bridalveil Falls appears at the
right, and in the distance Half Dome shows on the horizon.

National Park Service Photo

open out. The valley floor becomes wider and the steep cliffs have retreated. The road and stream both wander at will over a wide floor, and
the valley profile is obviously U-shaped. The feeling of scale here is curiously unreal. It is almost impossible to realize the height of the immense
cliffs soaring upward around us. Some of the early explorers in the region
experienced this same sensation of unreality, and they made estimates of
the height of El Capitan that varied from 400 feet to "at least 1,500 feet."

Actually El Capitan rears fully 3,600 feet above the valley floor, or about twice the height of the Rock of Gibraltar.

There are two especially fine vantage points from which the major features of interest can be seen. One is on the Wawona Road, as it climbs up the steep south side of the valley, and the other is at Glacier Point. From the Wawona Road we can get a view up the valley toward the east just before we pass through Wawona Tunnel. The massive cliff of El Capitan towers almost vertically from the valley floor on our left, while somewhat nearer us a large mass of landslide debris gives a curving slope to the side of the valley. On our right, Bridalveil Falls makes its exquisite leap of 620 feet from the end of a small hanging valley perched high up over the main floor. From where we stand we can see very clearly the trough-like depression of this hanging valley, and note how the far side rises into the rocky pinnacles of Cathedral Rocks. Such hanging valleys are common all over the world in connection with mountain glaciation. They owe their origin to the relatively rapid and deep carving of the main valley by a glacier which lowered the valley floor and increased its width so that it lies well below the level of entering tributary valleys.

Yosemite Valley has a U-shaped cross profile here, with the higher sides of the U standing almost vertical. In the distance, nine miles away, appearing just over the shoulder of the valley wall is Half Dome, of which we shall have a closer view later. This peak soars 4,800 feet above the floor of the valley.

By continuing up Wawona Road, first through the tunnel, then forking left at the Chinquapin Ranger Station to follow the Glacier Point Road, we eventually climb 3,000 feet to Glacier Point. From here we can look directly down on the wide flat floor of the main part of the valley and the meandering Merced River 3,200 feet below us, and we see much of what was hidden from our view at our first stop, lying as it did behind the large masses of Cathedral Rocks and El Capitan.

Across the valley and slightly to the left, Yosemite Creek in two major leaps and a series of very steep cascades makes a breath-taking drop of 2,435 feet from its hanging valley. In the spring when the snows are melting from the uplands and the rivers are in spate, this thunderous cataract, falling from a height of half a mile, is a most impressive sight. Later in the season the volume of water is greatly reduced in this and all the other falls of the valley, and in summer many may cease altogether. It is thus in early summer when its flow is diminishing that Bridalveil Falls lives up best to its name, as the winds in the valley blow the thin stream of water first one way then another into a slowly cascading, filmy veil.

To the east, Half Dome, now only about two and a half miles away, still towers over us and seems to dominate the scene with its insistent presence. From here at Glacier Point we can clearly see how the main part of the valley divides into two tributaries. The Tenaya Valley to the left of Half Dome is largely hidden from view, but to the right of Half Dome we have a wonderful view up the wide steep valley of the Merced and we can see how its floor drops in two major steps, which produce the exquisite Nevada Falls and then Vernal Falls farther downstream.

Further to the east, as if to put these beautiful and impressive valleys into their proper place, the High Sierras rise peak on peak to a jagged snow-covered horizon twenty miles away. In the summer this is a land of rounded granite domes of dazzling white. The explanation for the consistently rounded shapes lies in the peculiar type of weathering which has overtaken the granite here. The domes are literally peeling off in layers, rather like enormous onions.

In order to gain the proper perspective so that the origin of this scene

A before glaciation
B during glaciation
C after glaciation

YOSEMITE VALLEY, CALIFORNIA
after Matthes

can be appreciated, it is well to remind ourselves of certain relationships and observations. Yosemite Valley runs from east to west and is cut into the westward-facing slope of the Sierra Nevada Mountains. This range is the largest continuous mountain block in the United States. It is 400 miles long and 40 to 60 miles wide. The crest line which reaches a high point of 14,000 feet at Mount Whitney is not at the center of the range but lies near the eastern border. From the highest parts of the range, there-fore, the slope downward to the east is very rapid whereas to the west it is gentle. This range is a block of the earth's crust which has been tilted westward as the result of an uplift of many thousands of feet along a crack or fault on its eastern margin. Such uplift, well above sea level, inevitably brings weathering and erosion in its wake. This explains the presence of the granite now at the surface. Granite is formed at a great depth by the slow cooling and solidification of hot liquid rock, or magma, and to be exposed at the surface as it is here at Yosemite Park means that the tremendously thick mass of rocks which covered it at the time of its formation must have been removed.

In preglacial times the Merced River flowed down the westward-facing slope of the mountain in a V-shaped valley for its complete length. Yosemite and Bridalveil Creeks joined it as ordinary tributaries and did not fall as they do now into the valley of the Merced. Later, the mountain block rose and the westward tilt increased with the result that the Merced River, flowing as it did directly down the slope, started to cut a much deeper channel for itself. Its tributaries, however, which flowed across rather than directly down the increased slope, did not cut their channels as rapidly, and thus eventually found themselves left behind in such a fashion that they had to finish their journey to the Merced via a series of rapids and small falls.

Next came the Ice Age, and glaciers developed in the higher reaches of the Sierra Nevada and flowed down any available valleys. The Merced was such a valley, and the ice which flowed through it eroded both the sides and the floor of the valley with the result that it was somewhat straightened, widened, and carved into the typical U-shape. At the time of maximum glaciation we can picture Yosemite Valley as almost abrim with ice which extended down just about to El Portal before it melted away into a stream of milky white glacial water, full of finely pulverized mineral fragments. Such white glacial water reveals the tremendous abra-

THE MARCUS BAKER GLACIER IN THE ST. ELIAS RANGE OF ALASKA. The view shows typical jagged topography resulting from the erosion of valley glaciers. Notice the cirques, arêtes, horns, and U-shaped valleys. The medial moraines on the major ice tongues are especially obvious. *Photo by Bradford Washburn*

sive power of moving ice which with its grinding teeth of gravel and boulders literally turns bedrock into rock flour. We know that El Portal marked the limit of the glacier because here the U-shape of the valley changes into the unglaciated V-shape.

At that time in its history the main part of the Yosemite Valley had ice between 2,000 and 3,000 feet thick in it. The only landmark which would have been visible above the icy waste was probably the peak of Half Dome, as it stood between the two tributary glaciers which flowed down the Tenaya and Merced Valleys. These ice streams would have been an impressive sight as they coalesced to form one mighty ice river. Each glacier would have had a load of debris riding on its surface or frozen solidly into its icy recesses. Such material, being especially thick near the sides, formed what is called "lateral moraines." Below Half Dome two of these lateral moraines, one each from the Tenaya and Upper Merced glaciers, combined to form a medial moraine at the center of the main stream. This medial moraine would have appeared as a black streak on the lighter-colored ice, delicately following each turn of the glacier and thus emphasizing its flow pattern. Air views of some of the modern Alaskan glaciers show medial moraines to perfection, and they help us to visualize what Yosemite Valley must have looked like at that time. Some of the Alaskan glaciers have many such medial moraines, a new one being added each time a tributary glacier enters the main stream.

When the ice in Yosemite Valley eventually melted away it uncovered Bridalveil and Yosemite Creeks which then, however, instead of completing their journey to the Merced via the series of preglacial cascades, were forced to descend in flying leaps.

A later advance of the glacier reached slightly beyond El Capitan and Cathedral Rocks. This ice tongue left a low terminal moraine and dug out the valley to produce a slightly sinuous hollow, which became a finger-shaped lake when the ice melted away. Subsequently, the lake was filled by deltas built into it by the Tenaya and Merced Rivers, which explains the wide flat floor of this part of Yosemite Valley, a somewhat anomalous feature in an otherwise rugged mountainous area.

After studying Yosemite Valley, we can easily understand the origin of the fiords of the Alaskan and Scandinavian coasts. A fiord is simply a drowned, glacially produced U-shaped valley. If in imagination we allow the ocean to rise and cover California until there is a thousand feet of water in Yosemite Valley we would have a long narrow estuary facing westward, with deep water and precipitous cliffs—a perfect fiord.

In addition to glacial features the Yosemite region is also noted for the impressive and intriguing forms into which the granite has been shaped. There are steep, very extensive cliff faces unrelieved by any crack for great distances, and as has been already mentioned, there is a sweeping landscape of rounded knobs and domes of rock obviously shaped into their present configuration by the peeling off or exfoliation of concentric layers. The massive quality of the granite here is actually quite unusual. Most rocks, whether they be igneous, sedimentary, or metamorphic, are cut by rather closely spaced cracks or joints. The Yosemite granite does have such joints, but they are very widely spaced, sometimes many hundreds of feet apart.

Half Dome illustrates perfectly the development of both the cliff face and the rounded dome. The cliff on its northern side is due to one of the rare sets of vertical joints which allowed this face to be exposed after the collapse of half of the dome when glacial erosion in the valley below had undermined it. Actually, the location of all the major cliff faces in Yosemite Park were determined by such joint cracks. The straight cliff under Glacier Point, the face of El Capitan, and the cliffs under Yosemite and Bridalveil Falls are all joint surfaces.

The rounded southern part of Half Dome resulted from the process of exfoliation, and is due probably to a combination of three contributing factors. For one thing, the Yosemite granite is now under far less pressure than when it was initially formed deep in the earth. This decrease in pressure resulted in an outward expansion of the surface layers of rock with the consequent production of cracks parallel to the surface. Furthermore, the heating of the rock by the sun during the day and the cooling by night resulted in expansion followed by contraction which may have aided in forming the exfoliation cracks. It is known that heat from forest fires has caused the development of such cracks. Finally the chemical alteration or weathering of some of the minerals in the outermost layers resulted in a net increase in volume of the surface layers which then literally pried themselves loose from the fresher underlying material. Probably each of the three factors discussed had some part in developing the remarkable rounded domes of granite here, although there is some doubt as to the actual part played by the alternating heating and cooling of the surface layers.

The peeling layers vary in thickness from a few inches to a number of feet, as can be noted easily on practically any upland surface as one drives by. This type of weathering effect can be found at many places in the United States, especially where igneous rock is exposed. The granite at Franconia Notch in New Hampshire, incidentally, shows it very well.

The road westward over the crest of the Sierra Nevada through Tioga Pass and to Mono Lake passes over the High Sierra country, where it is much colder and the growth of trees and shrubs is not as lush as it is in the oasis-like Yosemite Valley. Many small lakes lie in the lower parts of the land here, and there are excellent views of a variety of glacial features as well as a multitude of rock knobs and exfoliation surfaces. Lake Tenaya is surrounded by white ledges of granite; in addition to their gleaming exfoliating surfaces they are locally enhanced by a fine, faintly striated polish which has almost a silky luster when looked at in certain lights. In places this glacially polished surface is peeling off to leave a rougher interior exposed, but there are still large areas of polished rock just as the glacier left them.

In the Sierra Nevada region there are still about sixty small glaciers which occupy the hollow floors of existing cirques. They appear very insignificant indeed in comparison with the Alaskan or Alpine glaciers, and in fact many may look much more like large piles of snow which have not quite melted than like true glaciers. They are glaciers, nevertheless, and may be distinguished from a pile of late melting snow by the basic glacial characteristics possessed by them. A true glacier persists throughout the year and consists of layered ice at the base which grades upward through granular snow to fresh snow at the top. And it has motion, which is indicated by the piling up of a moraine at its edge and by the development of a crevasse at the head end, next to the mountain.

AT THE present time valley glaciers are found in many of the higher mountains of the world. In North America they are especially well developed in the Alaska Range, as we have mentioned. Small glaciers in the form of ice caps occur on Baffin Island, Devon Island, and Ellesmere Island in the Arctic region of Canada. They are in essence small editions of continental glaciers, such as now cover only Greenland and Antarctica.

At the time of maximum glacial development, ice covered essentially all of Canada and extended well down into the United States. The southernmost extremity of the continental sheet followed roughly a line from New York City through Pennsylvania to the Ohio River, then down the Ohio, and up the Missouri River to the vicinity of Great Falls, Montana. From this point a line approximately one hundred miles south of the United States-Canada border to the Pacific divides the country to the north, which was almost entirely overridden by ice, from land to the south, in which the glaciers existed only at the higher elevations in the mountains. In contrast to the glaciers which covered the eastern two-thirds of North America, the ice which covered the mountainous parts of the West

consisted of many separate, individual masses. They were valley-type glaciers which were merely extensions of those still in existence today with
the addition of many more starting at lower elevations. Ice from these
many glaciers flowed down onto the plains at the foot of the hills and
coalesced with one another, producing extensive piedmont glaciers. The
continental ice sheet which flowed outward in all directions from nour-
ishing grounds in the area of Hudson Bay met these piedmont glaciers
in the plains of Alberta not many miles from the foothills of the Rocky
Mountains.

At such times of glaciation a great deal of the water of the world was
locked up on the land in the form of ice. As a consequence the sea level
must have been much lower than at present, perhaps as much as 300 feet.
This would have exposed a great stretch of the continental shelf along the
East Coast, making a shoreline many miles beyond today's coast. The very
lowest level of the sea would of course have coincided with the maximum
development of glaciers on the land. The recent postglacial drowning of
the coastline is due in large part to the return of ice water to the sea. If the
job were completed and all the present glaciers in the world were to melt,
the sea level would rise still further, probably between one and two hundred feet.

The enormous mass of ice, concentrated as it was in Canada, was many
thousands of feet thick, and its sheer weight on the earth's crust produced
another very interesting effect. It pushed the surface of the earth down by
as much as 1,000 feet where the load was greatest. Since the ice left, the
land has been rebounding in an attempt to attain its preglacial elevation.

The concept of such a postglacial uplifting of the land is somewhat
surprising, and one may well ask what evidence there is for it. The an-
swer lies in a whole sequence of obviously wave-cut cliffs and former sea

beaches far above the present-day sea level, in the Hudson Bay area of Canada. They were cut well before the land had completed its uplift. As a matter of fact, there is evidence that the land here has risen a number of feet in the last few hundred years. Similar evidence indicates an uplift in Western Canada near Vancouver of 750 feet, and a rise of over 900 feet in parts of Scandinavia. Further south, in both North America and Europe the rise has been progressively less, with no evidence of similar postglacial uplift beyond the margin of the former ice sheets.

There is a very striking correlation all over the world between lakes and the former presence of glaciers. The valley-type glacier leaves rock-basin lakes, or tarns, in the floors of cirques, and it digs elongated hollows to produce finger-shaped lakes. Terminal or lateral moraines of such glaciers may also in places dam up side-streams to produce lake basins. In the case of continental glaciers, most lakes are due to kettle-hole formation or the damming up of previous drainage channels by glacially dumped material. Minnesota is a "Land of Lakes" and Michigan is a "Water Wonderland" only because at one time ice covered these areas, and dumped tremendous quantities of material in what appears to be a hit-or-miss fashion. In New York State the glacier sent pioneering tongues of ice down a number of preglacial valleys, with the result that these avenues were widened and deepened and turned into elongated receptacles for the very beautiful Finger Lakes. Such U-shaped valleys are not commonly associated with continental glaciers, but where tongues of ice can reach out from the principal ice mass such features do result, as we found in New Hampshire.

In the past half billion years there have been only two times when ice flowed off the land instead of water, one about 200,000,000 years ago and the other in the immediate past in the Pleistocene epoch, which started about one million years ago. Up to the present there have been four major and many minor advances of the ice during this recent time of glaciation. Perhaps in the future there may be other advances, and we may well be only in an interglacial time. Predictions are very difficult and depend a great deal on what theory of the cause of glaciation one favors. At any rate, glaciation was the most recent geologic event of major importance to occur in North America, and has given, as we have seen, the final scenic frill to Canada and the northern part of the United States.

PLAINS NEAR MORLEY, IOWA *Fairchild Aerial Surveys, Inc.*

CHAPTER 4

THE landscape in the central part of the United States is unbelievably flat
to anyone who has grown up among hills and valleys and has never before
experienced the vast sweep of the land in this area. Mile after mile of
straight roads traverse gently rolling country. From the slight broad elevations, fields of corn and grain can be seen stretching as far as the horizon,
with here and there a clump of trees or a farmhouse to break the pattern of

The Plains

the fields. The sky dominates the whole, and cloud formations from fleecy white to threatening black are an important part of the scene. This is Iowa in the summer. It appears that the earth has been tamed to a common norm, and that the scenic elements consist entirely of the sky and the works of man, his buildings and the crops he has planted. The cultivated fields are laid out very meticulously in neat rectangles, a design obviously controlled by the road system. This rectangular pattern is strikingly consistent, as we can see at once if we glance at road maps of such states as the Dakotas, Nebraska, Kansas, and Iowa.

The roads go essentially in only one of two directions, north and south or east and west. In this flat country a road builder can set his sights on his goal and go directly there without having to worry about cliffs and gradients which are too steep. The pattern here is due to the way in which this part of the country was opened up to settlement. The United States Government sold rectangular parcels of land to the early settlers. Roads which were laid out between these areas thus determined the pattern we see to-

GLACIAL MAP OF THE UNITED STATES. Ice at one time or another covered all the area indicated, but not the whole area at any one time.

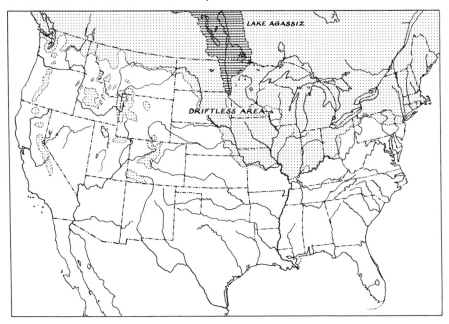

day. It can be noted that the road density varies widely. For instance, it is greater in Iowa and eastern Nebraska than in central and western Nebraska, reflecting the greater density of population where the rainfall is heavier.

The rich farmland of Iowa, set out in neat rectangles, is totally unlike the glaciated country we have already visited. Nevertheless, this region was at one time also covered by ice. A thin coating of till was spread over an unusually flat landscape, obscuring the bedrock and leaving the rounded, slightly rolling surface we see today. Elsewhere in the glaciated parts of the interior plains the preglacial landscape has been smothered more deeply by glacial debris or "drift," in places hundreds of feet thick. Extensive terminal and recessional moraines may be found stretching for miles across the land, and they often form the major relief features.

On a trip westward from the Appalachian Mountains of Pennsylvania, through Ohio, Indiana, and Illinois, we cross a number of these moraines. Many of them are composed of till quite unlike that of New England, and may not appear familiar. We cross a very good example of such a moraine near South Bend on the Indiana Turnpike. A hummocky ridge, a few tens of feet high, with no boulders or stone walls in evidence, and showing a clayey soil, stands in marked contrast with flat country on either side.

Moraines are composed of material available to the glacier, and if only fine-grained sediment such as shale is at hand the composition of the moraine will of necessity reflect this. Farther north, the moraines are likely to contain more boulders as they are nearer to a source of igneous and metamorphic rocks which outcrop in Canada and around Lake Superior.

We become accustomed to smooth, rolling country as we cross Ohio and Indiana, and we have small opportunity to observe rock structure. However, if we are traveling on Route 20 in western Illinois we enter a region with a dramatically different type of landscape shortly after leaving Stockton. In this area which extends about thirty miles to the Mississippi River at Dubuque, Iowa, bare layers of sedimentary rocks are very obvious on the sides of the ridges. This is a section of the "Driftless Area," a part of the country never covered by the ice which at one time or another was spread over all the surrounding area.

The Driftless Area is a region of about fifteen thousand square miles, largely in Wisconsin with a narrow wedge in northwestern Illinois, its western border following the Mississippi River. On entering this area from any direction, we are abruptly made aware of what the passage of a huge mass of ice has done to the land we have just left. We now come upon buttes and delicate rock pillars. Such irregular hills of bare bedrock would have been quickly destroyed and obscured by any overriding mass of ice.

This region was never covered by ice because of two complementary factors in the topography of the land. The resistant rocks of Northern Wisconsin form an area of highlands, and standing as they do between the lowlands of Lake Superior and Lake Michigan they formed a bastion, which during the Ice Age caused the channeling of the ice down the lowlands on either side. Thus in the lee of these highlands we are able to visit an area never covered by ice.

The Driftless Area occupies only a very small section in the over-all expanse of the glaciated country we cross as we continue westward. Approaching the Iowa-Nebraska border we find ourselves in a region where the landscape was produced indirectly by glaciation. Here we can see somewhat less obvious but locally very important results of the glacier's former presence in the thick and widespread deposits of loess. These are wind-deposited layers of dust which cover a great deal of the central part of the United States. They produce noticeable scenic features along river banks and road cuts where dust walls stand up as brownish vertical cliffs. Near Sioux City, Iowa, the Missouri River has cut down to expose loess bluffs, some of which are over 100 feet high. These are especially notable north of the city.

Cliffs made of dust may seem somewhat of an anomaly, but they can be found in many places in the world. Such deposits maintain their vertical stand because they consist of microscopic angular particles of quartz, feldspar, and calcite. These mineral fragments interlock and thus form a very stable three-dimensional network. Such a deposit is easily cut with a knife; in fact, caves carved in it will maintain their shape for long periods of time.

The origin of these loess deposits is intimately associated with the Ice Age. Much of the outwash material from any glacier consists of fine mud which on drying out can easily be blown around. Such mud deposits contain a great deal of unweathered material, produced by the abrasive action of a glacier as it ground over the bedrock. This explains the presence of the angular pieces of feldspar and calcite particles in the loess. Usually, in the normal course of weathering, feldspar is chemically changed into other minerals, and the calcite is dissolved and carried away in solution.

The loess cliffs of Sioux City are indeed an insistent reminder of the Ice Age, once we realize how they were formed. Directly north of us lies another region, utterly unlike the country around Sioux City, where once again we must bear in mind the former presence of a glacier if we are to appreciate it fully.

At Grand Forks, North Dakota, the main characteristic of the landscape is its flatness. Grain elevators stand like skyscrapers, and one looks in vain as far as the horizon for the slightest undulation in the surface. If we follow the main road westward we must travel forty miles, or about half way to Devil's Lake, before this country changes to a land of low rolling hills.

The explanation for this type of country, as for all really flat areas in the world, lies in the deposition of sediments rather than in erosion. In this case deposition took place in a lake now disappeared, Lake Agassiz.

The Red River which forms the boundary between North Dakota and Minnesota flows northward through Manitoba and eventually into the Arctic Ocean. Toward the end of the Pleistocene epoch this path was blocked in northern Canada by a barrier of ice. This temporary dam impounded Lake Agassiz, which at the time of its maximum development covered an area larger than all the present Great Lakes combined. It was drained when the ice eventually melted from northern Canada, and evidence of its former presence now consists of the extensive horizontal deposits of mud and sand which were washed into the lake and spread out on its floor.

An arm of Lake Agassiz extended southward from Grand Forks 150 miles. It was about 80 miles wide, and covered the present North Dakota-Minnesota boundary line. To the north and east the lake opened out in Canada to include the Lake of the Woods, Lake Winnipeg, and many other present-day lakes which now fill the lowest parts in the floor of the otherwise extinct lake.

Since the time of Lake Agassiz, streams have cut a few narrow and shallow valleys in the former lake floor. These are merely meandering grooves ten to twenty feet deep in an area which has otherwise remained largely undisturbed since it was drained.

Other glacial lakes left evidence of their presence in the form of flat-floored country, but none of them approached the size of Lake Agassiz. The present-day Great Lakes spilled over onto adjoining country at various times during the Ice Age. The former extent of such enlargements is shown primarily by areas of flat country which now fringe most of the lakes, and which in places may extend many miles away from present shorelines. Chicago is built on the floor of an extended Lake Michigan, and Detroit on a former floor of Lake Erie.

Continuing our trip westward from Grand Forks on U.S. Route 2, through the northern parts of North Dakota and Montana, an irregular surface can be noted almost all the way to the Rockies after Lake Agassiz has been left. Once again we are in a region of glacial drift. After a time of rain many ponds will be found filling the hollows and kettle holes in the stony morainal dump. This route follows a path very near the southern limit of the continental ice sheet, which in this area barely pushed across the United States-Canada border. South of the glacial limit, by contrast, there are no small lakes or any vestiges of haphazard morainal piles.

Except for the Driftless Area, all of the major landscape features we have discussed so far in the plains area are primarily the result of glacial deposition. Outside of the glaciated areas, however, the major relief features owe their origin to the erosion of more or less horizontal rock layers free from any mantle of glacial debris. A line of bluffs bordering a river valley incised into a more or less flat surface can be found many times in the plains of the United States. A typical series of such river bluffs is located at Chamberlin, South Dakota. Here the Missouri River had a narrow flood plain which is now covered by the end of the Fort Randall Reservoir. On leaving the river a wide belt of knobby, grass-covered dissected land must first be climbed before reaching the flat upland of the plains away from the river. The rise is a number of hundreds of feet in a few miles. The ramifying valleys which are cut into the bluff zone grow rapidly shallower and come to an end in the flat plain above. The zone of dissection here starts at the flooded river. At other places where the river possesses a wide floodplain, such as farther downstream at Elk Point, South Dakota, the line of bluffs starts well away from the stream. Obviously, with the passage of time the zone of dissection retreats away from a stream as the side tributary valleys are cut more deeply and the major stream in its meandering course first impinges on one side of the valley and then on the other.

We can frequently trace such a line of deeply dissected bluffs along the larger streams, and can thus observe the forces of erosion at work, as smaller streams in flowing down to join the larger cut back more and more deeply into the flat upland areas. Such a development of gullies explains why there is a cut-up zone of country bordering the major stream valleys, rather than just a straight line of cliffs.

In the western part of South Dakota the White River has cut a deep valley into a slightly rolling upland surface. Lines of cliffs or scarps are now retreating from the river as streams wash down their slopes, and in places the cliffs are now many miles from their original position next to the river.

The dissection of the scarps is here far more intricate than that of the Missouri River bluffs and produces what is called a "badland" type of scenery. Such topography develops best where the material is soft and easily eroded, is distinctly above the level of the major stream which drains the area, and where the climate is somewhat arid. In these Badlands of South Dakota the White River has cut many hundreds of feet into the horizontal sediments of the White River formation, composed largely of fine clay with a few thin sandstone layers and other more resistant parts. The color of the rocks as a whole is either dazzling white or light gray, with here and there some pink shades and, more rarely, some rather intense reds, yellows, purples, and greens, due to various compounds mixed with the white clay.

The slopes are very steep and erosion is so rapid that vegetation has not had a chance to gain the least foothold on them. The flat areas are covered with grass and scrubby vegetation. If visited in the spring and early summer the green grass and bright blue and yellow flowers on the flats make a striking contrast with the dazzling bare ground of the slopes.

About fifty miles west of the Badlands we approach The Black Hills. They are really an outlier of the Rocky Mountains, whose principal mass lies still farther west about 150 miles at the Bighorns. However, if we approach the Rockies along a route about 300 miles farther south, in central Colorado we would note perhaps the sharpest transition from the Plains to the mountains. On a good day we can see them rising abruptly above

BADLANDS, SOUTH DAKOTA. The horizontal bedding is dramatically emphasized by changes in the color of the various layers.

Courtesy Publicity Department, South Dakota State Highway Commission, Pierre

the plains from more than a hundred miles away, especially when the presence of snow on the peaks helps us to distinguish what might otherwise appear to be a cloud bank on the horizon.

In western Kansas along this southern route, we will find a clue to the geological history of the Great Plains in the sides of some of the stream valleys. Here we can see horizontal layers of sand and gravel, which tell us that the flatness of the plains in this region is due to deposition. As we continue westward into Colorado, we enter an area of broadly rolling plains which gradually rise in elevation until the Rockies are reached. Here the original level land due to deposition has been so deeply eroded that now none of the original surface remains.

The geologic history of the western part of the Great Plains is tied in closely with that of the Rocky Mountains. The present Rockies represent a second generation of mountains. They were first formed by a buckling and uplift of the crust at the end of the Mesozoic Era, more than 60 million years ago. After their uplift, they were broken down by the weather and much of the debris thus formed was washed eastward and spread out over the lower-lying plains as enormous coalescing outwash deposits. In this way a thick mass of material was spread out all along the foot of the Rockies. It was in the form of a wedge which thinned out toward the east after covering a stretch of country some hundreds of miles wide. Eventually the initial Rockies were worn down to a moderately flat surface of erosion, which toward the east changed into the much flatter surface of deposition as exemplified in those areas in Western Kansas which we have just crossed.

More recently, the whole Rocky Mountain region has been arched up with the natural corollary that erosion started again. The present more jagged nature of the Rockies is the result. There are now a few isolated remnants of the previous relatively flat erosional surface still persisting in the higher parts of the mountains. The thick layers of sediments toward the east were also uplifted and streams started to cut into them as well as into the Rockies themselves. Near the mountains the wedge of sediments has been eroded more deeply than it has farther away. This accounts for the dissected part of the plains near Denver, which changes to less eroded country in western Kansas and northern Texas.

The geologic history of the Interior Plains for the past 500 million years has been in general a quiet one. During the Paleozoic and much of the Mesozoic Era encroaching and retreating seas spread alternating layers of limestone, mud, and sand over the area. Then after the Rockies were formed at the end of the Mesozoic Era the thick wedge of sediments in the western parts of the plains was laid down.

The most recent event of major importance was the coming of the glaciers to the northern reaches of the plains. They brought in a cover of debris and molded the landscape into new forms.

At present the plains are undergoing erosion. The rocks are being weathered and the soil washed into a new sea far removed from the ancient central sea which at various times covered much of the interior of the country. Except for the wedge of superficial sediments deposited in the western plains, most of the rocks were laid down under the sea. Fossils of the marine life of the time may now be found in them. As some of the older sedimentary deposits of Kansas are washed away, remains of exotic reptiles of the latter part of the Mesozoic Era occasionally come to light. Geologists have found some remarkable fossils, such as the remains of plesiosaurs, fifty feet long with necks like that of the fabulous Loch Ness monster, and thirty-foot fish-like ichthyosaurs.

Stream action in cutting into the wedge of sediments in the Badlands has brought to view remains of more recent animals which lived and died and were buried in the growing pile of stream deposits coming from the initial range of the Rockies.

Such fossil remains help us to create in our minds a picture of the landscape of forty to fifty million years ago. This South Dakota country of the past was a land well watered by streams, with plenty of grass and vegetation, where small horses, camels, and rhinoceroses lived and grazed, and where such carnivores as tigers and wild dogs lived well on the rich game. There were swamps and mud flats in which the animals were caught and buried, and wide meandering rivers which continually shifted their courses, and in doing so built up an ever-thickening mass of sediments.

Everywhere in the interior plains we find sedimentary rocks. As we drive across this country we are continually aware of the horizontal structure of the land. Although there is considerable variety in the landscape, as we have seen, we are left with an impression of a region of low relief, of incredibly flat areas, gently undulating plains, low hills, and shallow valleys. The scenery is pleasing but on the whole its beauty lies in the serenity of its wide horizons and the sweep of the land rather than in any dramatic contrasts in the relief of the land itself.

MESA NEAR GALLUP, NEW MEXICO *New Mexico State Tourist Bureau*

CHAPTER 5

THE Colorado Plateau is a land of extremes. Forests contrast with deserts. Extensive flat areas alternate with immense tortuous canyons, cliffs, spires, mesas, and buttes.

The Grand Canyon of the Colorado River, in northern Arizona, which is the best known feature of the whole area, strikes the traveler with an impact that is immediate and unforgettable. However, it is difficult to travel

Plateau Country

anywhere in this region without finding dramatic scenery. There is Bryce Canyon, with its glowing, vivid rocks carved into delicate airy pinnacles; thrusting upward into an intensely blue sky, the flaming stone spires are almost unbelievable. Cloud shadows on the Painted Desert enhance a landscape already incredibly colorful, with a palette ranging from deep purple, through lavenders and pinks, to delicate shades of yellow and

PLATEAU STRUCTURE. The Colorado Plateau about forty miles south of Moab, Utah, near Grand View Point. Note the cliffs of resistant rock and slopes of weaker material.

Photo by Laurence Lowry

an infinite variety of brown earth tints. At Mesa Verde we find gems of ancient Indian architecture built beneath the shelter of huge overhanging sandstone cliffs. The rocks are streaked black from the smoke of long-dead fires of a vanished people. Monument Valley is a world littered with buttes and rock towers, the rugged survivors of a disappearing plateau. The day is short at the bottom of Zion Canyon; here white sandstone cliffs tower thousands of feet above a narrow canyon floor. Near Moab in eastern Utah there are many colorful stone arches, windows, and spires. And in southern Utah, Rainbow Bridge, carved in brilliant orange-yellow sandstone, takes a flying leap of nearly two hundred and eighty feet as it spans Bridge Creek, forming one of the most incredible sights in this land of variety.

Everywhere the basic horizontal layered structure of the land is obvious in the even skyline and in the arrangement of the rocks as seen on canyon walls and cliff faces. Rivers have sliced deeply into the sedimentary layers, so that we can now obtain in many places an excellent edge view of the structure. The cutting of the plateau into canyons has led to the formation of cliffs formed of the stronger rocks alternating with slopes of weaker, more crumbly material which is less able to stand by itself.

This is a paradise for the tourist, and here many geological phenomena and principles are beautifully illustrated. The amazing variety of scenic features is indeed astonishing when we realize that the same fundamental horizontal plateau structure is found everywhere.

The Colorado Plateau in all covers about one hundred and thirty thousand square miles in Utah, Arizona, New Mexico, and Colorado. At the higher elevations the rainfall is adequate for the development of huge forests, and in the lower and dryer regions one finds some of the grandest desert areas of the world.

In general, outside of the limited area represented by canyon floors, most of this region is over five thousand feet in elevation and parts of it rise to eleven thousand feet.

The Painted Desert of Arizona lies in one of the lower and dryer parts of the region. Here layers of soft crumbling shale with vividly bright colors have been uncovered by erosion. Iron-stained red, orange, and yellow rocks contrast with light gray and white layers. Since there is no vegetation to cover up the land the colored rocks stand out in all their rainbow brilliance in the bright sunshine. Petrified tree trunks are found in some of the layers; a few are now completely exposed to view looking like trunks of trees which have just been felled. The illusion that this is someone's axework is increased by the way in which the petrified wood has broken into small slivers looking like chips from a log, chips which are found to be no longer made of wood but of hard silica which through geologic time has replaced the wood.

The presence of these stone "wood piles" in today's landscape shows that the two contrasting events of burial followed by disinterment must have taken place since the time when a forest covered this area. Initially the trees were buried quickly by layers of mud and sand so that the wood did not rot or break up on the surface. While thus buried, percolating ground water brought silica, in solution, which replaced the woody fibers, layer by layer, so that even the growth rings of the trees were often preserved in the stony replacement. A forest turned to stone thus lay buried for millions of years, until uncovered by the weathering and erosion of the soil above it. Now we have the fallen forest back again, lacking the more delicate leaves and twigs, but with its branches and trunks of stone strewn over the present landscape.

Curiously enough the major agent of erosion which uncovered these

reminders of a once much moister climate is rain and stream wash. Obvious water-worn slopes and stream-cut gullies are apparent everywhere.
The major force molding the landscape here, as in many other desert
areas of the world, has been that of running water, brought by the in-

PETRIFIED WOOD WEATHERING OUT OF SHALE at Petrified Forest National Monument, Arizona. *National Park Service Photo*

frequent rains. The wind of course plays a role, but in general it is an unimportant one. It merely shifts loose material from one place to another, much as it does in building and shifting of sand dunes.

The horizontal sediments of the Colorado Plateau are composed of alternating layers of sandstones, shales, and limestones, altogether many thousands of feet thick. Embedded remains of shallow sea-dwelling animals clearly show that most of the layers were laid down on the floor of the ancient seas which at various times flooded this part of North America. It took over half a billion years for the thick sequence of sediments to be built. The deposition was not continuous throughout this time, as there are large gaps in the record, interruptions which show that for long periods, perhaps millions of years, the land stood above water. It was at such times that the trees of the fossil forest grew and died or winds piled desert sand into great dunes, similar to those whose solidified remains now stand exposed in Zion Park, or rivers wore an uneven surface into the land. Eventually, after each of these episodes, the sea again flooded the area until the tremendous mass of plateau sediments was formed.

After the time of sedimentation large-scale diastrophic forces pushed the layers straight up in the air, up where the agents of weathering and erosion could attack them. Wherever the wearing-down processes concentrated their force canyons are now found, and the most superb of these is the Grand Canyon of the Colorado River.

The drive to the North Rim of the Grand Canyon involves a climb of many thousands of feet, for it is on top of the Kaibab Plateau, a part of the major Colorado Plateau that has been uplifted somewhat more than its surroundings. From the north the road through Kanab in southern Utah crosses very shortly into Arizona. A long slow climb takes us from the dry countryside of the border region into a land of magnificent forests of Ponderosa pines, once the summit of the Kaibab Plateau has been reached. From Jacob's Lake south for forty miles to the actual rim of the Canyon, the road runs through a flat country of beautiful woods alternating with stretches of parkland. Driving through this area on a summer evening, we may see hundreds of deer grazing peacefully, and we have to drive with care to avoid the small fawns that leap across the road and dash into the brush on either side.

Then, standing on the brink of the North Rim we seem to be on top of the world. The chasm of the Canyon opens out, and the South Rim can be seen in the distance. Beyond it, the flat surface of the plateau extends to the horizon, with here and there small hills rising above it. These little hills are recent volcanoes, formed when lava came up through the plateau rocks and spilled out on the surface. From our vantage point we are actually looking down at the South Rim, twelve hundred feet lower in elevation. This disparity in height makes a tremendous difference in the amount of precipitation in the two areas. On the North Rim we are in a great forest land with rich grazing pastures whereas on the South Rim only scrubby, semiarid vegetation can survive. And, because of the heavy snows that fall on the northern side, the visiting season must end about the middle of October, yet the South Rim is open all year around.

The South Rim is about twelve miles away. On a road map the traveler may be warned by a notice that reads: "Travel across the Canyon by foot, muleback, or airplane only." By car the journey to the South Rim is two hundred and fifteen miles. The Kaibab Plateau must first be descended and the Colorado River crossed at a spot where the canyon is narrow enough for a bridge to be built across it. This happens to be just a few miles from Lee's Ferry, which was the only place for hundreds of miles where the river could be crossed in the early pioneer days. Leaving the river, the road to the South Rim next traverses part of the Painted Desert and then climbs the plateau once more, locally on this side of the canyon called the Coconino Plateau.

Most people, on actually standing on the South Rim of the Grand Canyon and looking out and down into the abyss, find that they are totally

THE GRAND CANYON OF THE COLORADO RIVER, ARIZONA, from Lipan Point on the South Rim. The even skyline, horizontal bedding, and the alternation of steep cliffs and gentle slopes are the characteristic elements of most scenes in the Colorado Plateau area. *Photo by Josef Muench*

unprepared for the grandeur and immensity of the scene. It takes time and imagination to grasp the parts which compose the whole. Here on the South Rim we are below the level of the opposite North Rim, and for this reason the Canyon seems a little more overpowering, since we have the feeling of being already in it. Also, the river itself is closer to us, and from a number of vantage points we can look almost directly down six thousand feet and see its watery thread glinting in the sun. On a clear day the Painted Desert can be seen forty miles away to the east.

The day's trip to the bottom of the canyon by muleback is a truly rewarding experience. The aches and pains connected with this mode of transportation are soon forgotten, but the many wonderful views and the novelty of descending a mile straight down into the crust of the earth are long remembered.

It takes from three to four hours for the mules to pick their way down the narrow winding trail to the bottom—eight miles of travel to descend one mile. Another world exists down here. Climatically, we might be in Mexico, many hundreds of miles to the south. The air is warm; a few hundred yards up Bright Angel Creek, the little oasis of Phantom Ranch seems almost tropical in its lush greenery.

Here at the bottom of the Canyon the brown and turbulent Colorado is the most insistent part of the scenery. This churning river as it moves toward the sea carries an immense load of mud, sand, and gravel, some of it in suspension and some of it rolling and scraping along the bottom.

It is not surprising to anyone who has once seen this turgid, sediment-laden river that water running off the land is the greatest transporting agent in existence today. The wind, the ocean, and the glaciers all combined carry but a fraction of the immense tonnage of rock debris that is moved annually by rain water as it returns to the sea. Carefully listening to the roar of the river we detect that it is compounded of many different sounds. Behind the surface gurgle and splash one can discern the deeper tones made by boulders in transit as they grind on each other and abrade the bedrock of the canyon floor. This is indeed forceful evidence that this youthful river is continuously striving to cut deeper as well as to carry its gargantuan load of debris. It has been estimated that a solid boulder of hard rock six feet in diameter would be entirely reduced to mud and sand if it were subjected to this kind of battering for only four years.

As one becomes accustomed to the roar of the river, other aspects of the scene can be appreciated. The rock walls down by the river, unlike most of the canyon higher up, are no longer in clean-cut horizontal layers, but in places appear to be composed of a gray unlayered crystalline material and in others of a gray resistant rock with almost vertical banding. The first is a granite and the second a schist. They bear witness to a time of mountain-building and volcanic activity many millions of years ago. The granite was intruded into the schist as a hot liquid and then solidified, and the schist was produced when some previous rocks, perhaps sandstones and shales, were heated and squeezed; the sedimentary characteristics of these original rocks have been utterly destroyed by the heat and pressure.

The Colorado River at this point has actually cut through the plateau and uncovered to view the basement rocks on which the plateau sediments were originally laid. Here in the Granite Gorge we are walking on some of the most ancient rocks in the world, well over a billion years old. They represent the roots of mountains which existed and were worn down long before the plateau rocks were deposited.

These basement rocks, on closer examination, are found to consist of two distinctly different groups. Here and there we can detect remnants of a series of rocks which overlie the granite and schist and underlie the horizontal layers of the plateau proper. These patches are seen to be layered sedimentary rocks tilted at a slight angle. Thus we now know that the lowest part of the Grand Canyon is composed of two sequences of rocks: the granite and schist, and a series of tilted sediments which were laid down, deformed, and eroded long before the main mass of the plateau sediments were deposited.

These two groups of rocks appear insignificant now, but the implications of their presence are stupendous. Their existence is evidence of two complete ages of mountain-building followed by intervals of time long enough for the hard rocks of the high mountains to be weathered and washed away until nothing but the truncated roots were left. This means that many thousands of feet of rock, representing millions upon millions of tons of material, must have been removed. The pace of this process of weathering and erosion was very, very slow, by comparison with the speed with which human events are accomplished. That a whole mountain range was worn away is evident in the nature of the rocks themselves. Granite and schist are both of necessity formed deep within the crust of the earth. Thus, we know that the high mountains in whose depths these "plutonic" rocks were formed must have been worn down to their very core before the sediments of the tilted sequence could be laid directly on the granite and schist.

The second chapter of the Canyon's history as shown by the patches of the tilted sequence of sediments consisted first of sedimentation, then uplift with tilting, then erosion, and finally subsidence of the crust to permit the deposition of the horizontal rocks of the plateau proper.

And now we happen to be present at a moment in earth history when it is possible to see the sequence of rocks from the third chapter in the canyon's history in the process of removal. The marvelous scenery of the present makes us wonder what awesome mountain scenery we may have missed in the past.

On the return journey to the rim of the canyon one has plenty of time to observe the entire sequence of rocks which have been exposed to view by the action of the Colorado River as it has cut backwards through the immense reaches of time, first through the plateau and then into the basement rocks. The trail mounts steeply, then flattens out, only to climb in sharp zigzags once more, and this is repeated many times on the way up. The color of the rocks changes, as well as the type, as we pass from zone to zone. The crumbly shale is evident on the slopes and massive limestones or sandstones at the cliffs. By this time it is obvious that the steplike valley walls of the canyon are due to these alternating layers of more resistant and less resistant rocks.

The first wide, gently sloping bench after leaving the narrow confines of the steep-walled Granite Gorge has been cut into a thick sequence of relatively weak shaly beds. This so-called Tonto Platform is a very obvious feature of the Canyon, and can be seen from any vantage point on the Rims. The ground slopes gently to form a terrace that is as much as a mile wide in places. Rising vertically above this broad slope is a massive cliff of red limestone—the Redwall. A freshly broken piece of this limestone, however, is white; the red iron stain on the surface has been washed down from the crumbling red sandstones of the Supai formation which lies directly over the Redwall. Above the Supai are two cliff-making rock formations, first, the Coconino sandstone, which stands out as a startlingly white band on the upper wall of the Canyon, and finally at the top the gray-white Kaibab limestone which directly underlies the surface of the ground at both North and South Rims. Probably the most conspicuous layers of the Canyon are the Tonto slope and the Redwall cliff, each of which can be traced as far as the eye can see, zigzagging into the distance, disappearing when followed up a lateral chasm only to reappear on the opposite wall, as well as encircling the various large erosional remnants left behind by the retreating valley walls. Shiva's Temple and Wotan's Throne are two of the largest of these buttes, many of which are visible from the vantage points on the rim.

After gaining the rim again the traveler emerges with a new respect for the magnitude of the job done by the Colorado River. A view back from the South Rim shows lengthening shadows in the deeper portions of the Canyon just visited, and a misty vista of the distant North Rim. Somewhere toward the darkening east we may see a local shower, its slanting gray lines of rain under a patch of dark clouds. It is at such a time that the Canyon seems eternal. But then one may note little, almost insignificant, signs of instability. A fragment of rock crumbles and falls from a cliff face nearby, the fall perhaps started by some small scurrying animal; a gust of wind blows a bit of dust or sand from the canyon face to fall further down the slope; or a trickle of muddy water flows down the rocks after a sudden shower. These are the ways that the canyon is widened. It is a slow, very slow process, but given time measured in thousands or millions of years it is absolutely irresistible.

Leaving the Grand Canyon we can drive north into Utah to Zion Canyon and then to Bryce Canyon; in so doing we will climb some of the most remarkable cliffs in this whole area. Furthermore, if we have been energetic enough to have made the trip to the bottom of Grand Canyon, by the time we leave Bryce, we will have crossed a sequence of rocks covering a span of more than a billion years. With a mental picture of the fundamental structure of this part of the land, it is a simple matter to see how this sequence is revealed.

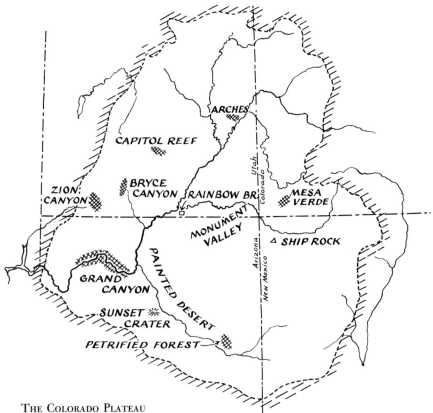

THE COLORADO PLATEAU

In the southwestern part of the Colorado Plateau where the Grand Canyon has been cut, the layered rocks were pushed up to form a broad arched
structure in which the layers dip very gently toward both the north and the
south. The crest of the arch actually coincides with the course of the Colorado River. As a result of this uplift the forces of weathering and erosion
have been given a chance to strip away the younger rocks which at one
time overlaid the present-day Grand Canyon rocks, and to allow the Colo-

rado River, as we have seen, to cut through the plateau altogether and uncover the basement material at the bottom of the Canyon.

To the south the flank of the arched structure is broken where the plateau ends, and the crust has been faulted and twisted in a part of the Great
Basin area. On the north flank of the arch, however, in southern Utah, the
tilted structure enables one to encounter younger and younger rocks as
one goes northward. The eroded edges of these northward-sloping layers
are now standing as cliffs facing to the south, like the risers of a giant stairway in which the wide treads are tilted slightly backwards, that is, to the
north. As a result of this tilt the top of each successive cliff is generally no
higher in elevation than the rim of the Grand Canyon itself.

On a trip to Bryce Canyon three steps of the stairway must be climbed,
the Vermilion, the White, and the Pink Cliffs. Each cliff line marks the
edge of a layer which is being peeled off the land. The Pink Cliffs represent the edge of the youngest or topmost layer, and are found furthest
north, where Bryce Canyon is formed by a deep reentrant cutting into the
cliff edge. Next older, and underlying the Pink rocks, are the White rocks
into which Zion Canyon is cut. Farther south, just north of the Grand
Canyon, are the Vermilion Cliffs. The variegated landscape of the Painted
Desert is formed of the lower layers of the rocks of the Vermilion Cliffs.

Such cliff faces are generally very far from straight, due to their irregular retreat under the forces of weathering and erosion. In places remnants
of the retreating plateau layers have been isolated and now stand out as
small mesas and buttes in front of the main line of cliffs. Curiously enough,
the origin of such features in this arid land is very similar to that of the sea
stacks off the Oregon and California coasts, which were at one time a part
of the mainland and are now isolated owing to the retreat of the sea cliff.

A landscape consisting essentially of nothing but such erosional remnants is centered in Monument Valley, in the border country between Utah and Arizona in the eastern parts of these states. Here the basic plateau structure is especially obvious in the flat tops of many of the "monuments," and in the stratification lines which encircle them. The scenery here is a transition between the flat plateau at a high elevation trenched with a few widely spaced canyons, such as the Kaibab Plateau, and the low, moderately flat region such as the Painted Desert, which is at the end of a long cycle of erosion.

A noticeable feature of erosion in an arid dry climate, such as that prevailing in Monument Valley, is the relatively small amount of talus or broken-up material at the bases of the cliffs. Apparently the talus material is broken into sand and dust fragments about as fast as it forms, and is washed away by the infrequent rains.

Returning to our "giant stairway" north of the Grand Canyon, we find that the plateau scenery at Zion Canyon has two features developed to perfection. Here a deep canyon has been incised into a thick layer of massive white sandstone. As one stands on the canyon floor looking upward, the sheer cliff seems to rise almost indefinitely. There are very few breaks in the vertical wall, and the influence of either joints or stratification seems to be at a minimum here.

A special scenic attraction of Zion Canyon is the beautifully developed "crossbedding" in some of the sandstones. This is a rather curious appearance of some of the layers which display a most distinctive line pattern. It is the result of the way in which the sand was originally deposited. Here we are looking at a cross section of solidified sand dunes piled up one on top of another. The lines of stratification dip first in one direction and then in another, showing the shapes of the dunes at various stages of their growth.

MONUMENT VALLEY ON THE ARIZONA-UTAH BORDER. Sandstone remnants above slopes of weaker sandstone and shale layers are all that remain of a once continuous layer.

Photo by Josef Muench

If we study such crossbedding it is possible to see that each set of layers was cut off at the top before the next layer was deposited. This truncation marks the partial removal of one sand dune before a new dune was piled on top of the old one. This strangely intriguing type of sandstone outcrop can also be produced in a somewhat more regular fashion by water-deposited material. The general sequence of events, however, must be the same, that is, the deposition of a series of layers on some sloping surface, such as the front of a growing underwater delta, followed by the partial removal of the old layers before a new series is deposited on them. Here in Zion the sandstone layers stand revealed in sheer bright cliffs, ranging in color from the deep jasper red of the Temple of Sinawava, through orange and pale pink, to the dazzling white summit of the Great White Throne.

At Bryce Canyon we can see a delicate, more intimate aspect of the plateau country. Here a cliff face hundreds of feet high has been eroded into a fantastically colored network of airy pillars, battlements, and ridges, with an infinite number of gullies, deep reentrants, and isolated erosional remnants. The early morning is an excellent time to visit this eastward-facing wonderland of ramifying gullies. The deep blue sky and dark green pines and junipers contrast dramatically with the intricately carved, predominantly pink rocks, stained with iron oxide. In bright sunshine the colored rocks reflect the light with a dazzling brilliance, and even in the shadows they seem to be aglow with an inner fire.

The landscape here is due to the weathering and erosion, primarily by water, of a moderately thick series of poorly consolidated red shales, with a few interbedded, more resistant limy and sandy layers. These more resistant layers stand out like beads on the slender pink pillars. The vertical development of the spires and pillars of the canyon has been controlled by a network of upright cracks or joints, which allow the agents of rock decay to work well down into the layered rocks, and which also guide the formation of the gullies. The vertical, joint-controlled part of the scenery is so obvious that at first it is hard to realize that Bryce is cut into part of the same vast level series of sediments that make up all of the Colorado Plateau. The horizontal line of the sediments can, however, be picked up by

BRYCE CANYON, UTAH. The erosion of horizontal sedimentary layers of varying resistance cut by a series of vertical joints has resulted in this fantastic landscape.

National Park Service Photo

noting the color changes in the sedimentary layers, as well as by the alternating weak and resistant bands. Many of the pillars are capped by a small
piece of the harder material, which has thus for a time protected the softer
underlying rock. Such a feature resembles in many respects the ice pillars
found on the surface of a glacier where a small boulder has protected a
mass of the underlying ice from the sun and rain.

The sediments at Bryce are much younger than those at Zion and the Grand Canyon, and also were never buried so deeply. Largely for these reasons they are not now so firmly cemented as the older sediments, and this in turn has resulted in the formation of such truly fantastic erosional remnants as the Queen's Garden, the Cathedral, or the Silent City.

The scenery of Bryce Canyon is thus the result of delicate selective sculpture by rains and ground water solution, guided by joint and stratification planes. In many respects it is similar to the badland topography of South Dakota, with infrequent heavy rains, scanty vegetation, and soft rocks.

Yet another interesting modification of the basic plateau scenery is to be found at Mesa Verde in southwestern Colorado. Here a large erosional remnant some hundreds of square miles in area has been isolated to form a mesa. But unlike most mesas, which have horizontal tops, this one is assymetric and slopes gently southward. The structure resembles in essence a floating raft, tilted and almost awash at one end, while the other rides high above the water.

The road to the top climbs the high northern cliff in a series of steep hairpin turns. Once the lookout point on the crest is reached, the very dramatic difference between the gradual slope southward, determined by the gentle dip of the resistant cap rock of the mesa, and the precipitous north slope is revealed.

From the crest rain water flows swiftly down the face of the north cliff and much less quickly down the south slope. As a consequence, the north slope is being rapidly worn back, while the streams flowing south are slowly cutting many parallel canyons. The mesa, which at first glance appears to be a complete whole, is riddled with canyons so that only about half of it remains. Mesa Verde in time will become still further cut up by these ramifying canyons and gullies and should eventually resemble somewhat the landscape of erosional remnants at Monument Valley.

At Mesa Verde the cap rock is a resistant sandstone. In the canyon walls there are many places where this rock has been undercut so that shallow caves have been formed. Beautifully preserved Indian cliff dwellings which have been vacant for over seven hundred years, are now found in these caves. The buildings stand silent and deserted looking out over the canyons whose flanks are dotted with scrub oak and pinyon pines. A visit to this area is enriching from both geological and archaeological standpoints.

In an almost inaccessible spot in southern Utah there stands a type of monument which is very rarely produced in nature and which adds greatly to the already rich variety of scenes in this plateau area. This is Rainbow Bridge. When the visitor stands beneath the soaring arch of stone, inevitably the question arises, "How did it happen?"

This bridge was produced by the erosional activity of the stream it now spans which originally had a meandering course over the floor of a valley. Conditions changed and it incised its channel into an underlying layer of pink sandstone. Two of the meander loops approached so close together that the stream was able to cut through the separating cliff of rock and thus produce a natural bridge. The stream then took the shorter course under the bridge and still flows there. The bridge was probably cut at a time when a great deal more water flowed down the channel than flows now. The present delicate shape resulted from the narrowing down by weathering of a much thicker bridge. Frost and rain, heat and cold are still active, and it is probably only a matter of a few thousand years before the bridge is destroyed.

Natural arches may also be formed by differential weathering, that is, by the selective decay and breakup of the slightly weaker parts of a mass of rock. The Arches National Monument near Moab, Utah, has many such features. Here a massive red sandstone is cut by a series of vertical master joints about twenty feet apart, which have controlled and caused the development of ridge-like erosional remnants sometimes over a hundred feet high. It is in these more or less isolated remnants that arches have been etched. Some of them have been formed entirely in this massive sandstone, while others such as Double Arch have been cut partly into it and partly into an underlying, closely layered rock. Such quarrying is due largely to water which enters small cracks in the rock and removes the cement hold-

DOUBLE ARCH, ARCHES NATIONAL MONUMENT, UTAH. *National Park Service Photo*

ing the sand grains together, with the result that the surface layers after being loosened flake off.

It must be apparent by now that the various parts of the Colorado Plateau show different stages in the inevitable cycle of change through which an uplifted sequence of horizontal sedimentary rocks must pass. The major part of the Plateau as a whole is youthful, that is, there are many flat areas at a high elevation which have just started to be cut by a few widely spaced canyons. The land is high enough and the rivers have a short enough path to the sea so that deep valleys or canyons can be carved out and a plateau produced. This by definition is an area of horizontal rocks, standing thousands of feet high, with valleys cut thousands of feet below the general level of the land. A plain has the same horizontal structure, but the relief, or the difference in elevation between the high and low parts, is only a few hundred feet.

The Colorado Plateau is now a land of decay and erosion. The flat-topped mesa, the pointed spire of the butte, the cliff, and the steep parts of the canyon walls are all due to the partial removal of formerly continuous resistant layers, which alternate with the less resistant material which now forms the sloping land below the cliffs. It should be emphasized that the basic theme of this country, as was true also of the plains but with less relief, is cliff and slope and flat skyline.

In order to picture what this country looked like in the past we must in imagination put back on the cliffs the material which has fallen from them and is now found at their feet as slopes of talus or broken debris, as well as all the material removed from the plateau by rainwash and streams. This would mean filling in the canyons and extending the eroded parts of the cliffs and mesas, thus producing a land showing but little scenic variety, a flat country high above sea level.

CAPITOL REEF NATIONAL MONUMENT, UTAH. A resistant sandstone forms the upper cliffs which tower over the slopes cut into weaker shales. *Photo by Josef Muench*

The future history of this area will be one of continued deepening and widening of existing canyons, the development and growth of tributary canyons, and the wearing back of the present cliffs and slopes. Thinking far enough into the future and giving the forces of decay and erosion time enough, we see that the Colorado Plateau has the ultimate fate of being reduced to a flat surface, somewhat resembling its initial stage but this time near sea level. No more will there be crags, cliffs, and canyons, but rather an area of very gentle slopes and sluggish streams. This will take time which must be measured in millions of years. But who knows but by that time the unpredictable mountain-building forces will again elevate this part of the earth; then the agents of weathering and erosion would start all over again patiently cutting out new canyons and carving new cliffs and slopes.

VOLCANIC NECK, SHIP ROCK, NEW MEXICO   *Spence Air Photos*

CHAPTER 6

THE Columbia Plateau, with a lava coating many thousands of feet thick, stretches for over 100,000 square miles, from the Grand Tetons of Wyoming on the east to the Cascade Range on the west. Here in southern Idaho, eastern Oregon, and southeastern Washington the volcanic "fires" have ceased so recently that many features appear as if they had just been formed. The plateau in conjunction with the volcanic cones of the Cas-

Volcanic Landscapes

cades forms the largest area of recent volcanic activity in the United States. It is thus an excellent place to investigate volcanic scenery; the only major feature which this area lacks is a currently functioning volcano.

A number of the cones in the Cascades showed slight activity in the last century. They threw out a few ashes, and a little smoke was seen to arise from them. In the present century only Mt. Lassen in northern California has been active. From 1914 to 1921 it showed a mild form of activity which reached a climax in 1915. In that year there occurred some rather abrupt explosions at the top, one of which demolished a stretch of forest by a blast of debris which "sand blasted" the trees. This now forms the Devastated Area. The trees were stripped of bark and pushed over by the blast so that all lay neatly pointing away from the top of the mountain. Some of the volcanic particles were driven into the tree trunks over an inch by the force of this explosion.

The formation of the Plateau itself began about 50 million years ago, when liquid rock started to emerge at the surface in the vicinity of the Cascades. The lava welled out of many cracks in the crust, in one flood after another, and spread over and filled in a very rough terrain, so that now, after almost continuous volcanic activity extending over 50 million years, a generally flat surface has been built up by the accumulation in places of thousands of feet of flows. Vulcanism has persisted up to within a few hundred years of the present, and we may very well be in a geologically brief period of reduced activity, which may possibly resume in the future.

At places in the plateau country the liquid lava obviously flowed with great ease, and flooded up into the mouths of preexisting mountain valleys to a distance of five to six miles, thus turning ridges into promontories and hills into islands in a "sea" of liquid rock. The aptness of considering the lava as a flood is very striking in many places.

In the eastern part of the Columbia Plateau in southeastern Idaho, U.S. Route 36 traverses the Snake River Lava Plain for many miles. Here is volcanic scenery, some of it formed as recently as 1,500 years ago.

Volcanic activity usually leaves a somber black world. At the Craters of the Moon National Monument, for example, there is a dark lunar landscape of cinders, ash cones, and jagged flows. This area of black desolation, near

Arco, Idaho, is only a little over 15 square miles in extent, but it has a tremendous variety of features due entirely to recent volcanic activity. It is a miniature paradise for the volcanically inclined tourist, as many features can be seen and studied which elsewhere are spread over a much greater area. The landscape consists entirely of a flood of black, twisted, broken, frothy lava, above which protrude a number of small cones of ash and cinders. The land is dry and seems at first sight to have very little vegetation

CRATERS OF THE MOON NATIONAL MONUMENT, IDAHO. A cinder cone is shown in the background and the edge of a pahoehoe flow in the foreground. *Photo by John Shimer*

cover. In some of the protected areas, however, such as the ash fields on the slopes of the cinder cones, small limber pines and aspen have taken root, and in the spring and early summer many small brightly colored flowers dot the landscape. On a bright warm day in such a season the air is pungent with sage, and butterflies can be seen and the tracks of deer noted. Life is thus beginning to invade this wasteland. But the most notable feature of the area is still the black desolation.

From the parking lot at the south end of the Big Craters, a view of a number of contrasting scenes can be obtained. To the southeast directly in front of us a number of "spatter" cones lie in a row. The lower parts of their slopes are cloaked with a mass of cindery rubble; above this there is a cover of reddish-black cindery lava. Such spatter cones are smaller than the cinder cones and far less regular in outline. They were formed by gouts of liquid rock which landed in a pasty condition and were thus able to adhere together and so build up cones of fragments. They lie along a major fissure in the area and probably mark the last stage of a fissure eruption which helped to build up the top layers in this part of the Plateau. Each of these spatter cones has a small opening at the top, only a few feet across, which can be seen to enlarge slightly with depth.

Slightly farther off to the east the smooth rounded convex slope of Inferno Cone shows against the sky. It is composed essentially of fine black cinders with here and there a tear-shaped volcanic bomb on the surface.

Behind us, three separate cinder cones have grown together to form Big Craters, a somewhat elongated pile of loosely packed material. The largest of these cones is about five hundred feet high with a crater about one hundred feet deep. There are a number of other cinder cones in the Monument, the largest being Big Cinder Butte which rises about 800 feet above the general level of the flows. The cinders of all were produced by the explosive churning action in the throat of the cone as the liquid frothy fragments of lava were blown out.

The volcanic bombs which lie scattered over the surface of the cinder cones range in size from a few inches up to a foot or more long. They were blown out as small gobs of liquid lava froth and froze into the typical elongated tear shape of bombs while in transit in the air. An exhibit at the entrance to the Monument has a variety of these bombs on view, as well as examples of the different kinds of lava found here. In addition to the typical tear shape, the bombs may at times assume an elongated twisted ribbonlike form, or such a bomb may turn back on itself, taking a boomerang shape. All the volcanic bombs, however, show the rounded contours taken by the liquid as it soared through the air and are composed of the black frothy lava called "scoria." In places flattened bombs have been found. They landed while still in a pasty condition and on impact assumed a pancakelike shape.

If a visit to the Craters of the Moon at the time of volcanic activity had been possible, it would have revealed a very busy scene. Liquids and solids were coming out of holes and cracks in the ground and gouts of pasty lava were spouting out of some of the spatter cones; the ground was shaking; there were the sulfurous smell of escaping gases and noise of falling cinders and large bombs blown out of the throats of the cones; from various vents steam could be seen and heard escaping under pressure. At night incandescent bombs emerging from the cones shone against the dark sky; they shot up into the air and then rolled down the sides of the cinder cones leaving a fiery trail of sparks. The advancing front of the lava flows, which during the day appeared black because of the crust of solid lava formed on its surface, glowed in the dark of night as the dull red of the underlying liquid shone through the breaking crust.

Today, when we look more closely at the sea of black lava surrounding the spatter cones we find that it has in general a frothy appearance and is composed essentially of the same material as the cinders and bombs of the cones themselves. Volcanic material often has a porous texture, due to the presence of dissolved gases in the lava. When the lava emerges from underground such gases are given a chance to expand, with the result that the liquid becomes a froth before it hardens. A somewhat apt analogy can be made with a bottle of ginger ale, which, with the pressure removed on being opened, allows the previously invisible dissolved carbon dioxide gas to escape in streams of bubbles. The dissolved gas in the lava is also the driving force causing the explosive bursts of so many volcanoes and lava flows. When the gas content of a "magma," the name applied to lava before it emerges, is low, a less porous kind of igneous rock results, and if the gas content is negligible, the flow commonly freezes into a very fine-grained rock, which is called basalt when black in color.

The surface of a flow usually crusts over while the liquid material underneath continues to move, causing the hard surface to be broken up into a sea of blocks, somewhat as ice breaks up on a river in the spring. Whenever lava starts to solidify in this fashion, one of two contrasting textures results. The Craters of the Moon has good examples of both types. The "aa" (ä-ä) lava has a rough jagged surface of pointed angular fragments in great disarray, formed by the cracking of the hardened surface by the continuous drag of the molten material underneath. In striking contrast, the "pahoehoe" (pä-hō-ā-hō-ā) lava has smooth rounded surfaces and appears as twisted, folded pleats and wrinkles of lava. In places it looks very much like twisted taffy which has oozed and bulged. Some of the ropelike cylinders of pahoehoe are over a foot in diameter, and resemble logs of petrified wood. Its crust often has a barklike appearance, and there are cross frac-

CRATERS OF THE MOON NATIONAL MONUMENT, IDAHO. Details of pahoehoe lava and its flow pattern after removal of the surface crust. *Photo by John Shimer*

tures which are very similar to those found in fossil wood. If the crust, which is generally less than one inch thick, is removed the underlying lava is seen to have had a life of its own, often having flowed in a direction different from that of the surface material, producing some beautiful patterns of stringer and knotlike formations. This underlying lava obviously was squeezed and continued to flow after the surface had frozen. Sometimes there is a glossy dark iridescent veneer on the pahoehoe flows. This is especially noticeable on the Blue Dragon Flow toward the south.

Both the aa and pahoehoe types of flow surface are formed when motion continues after partial solidification of the lava or at least after it has reached a pasty condition. Pahoehoe is formed when there remains in the lava plenty of dissolved gas but little agitation of the lava; thus, the gas is not lost, and a plastic mobility is maintained. Aa, on the other hand, loses its gas content quickly and, hence, its mobility. In Hawaii some recent flows of pahoehoe near their source have been found to change to aa farther down the slope.

The fluidity of any magma depends primarily on the presence of dissolved gases. For example, in Hawaii the temperature of a liquid basalt lava with some dissolved gases may be as low as 850 degrees centigrade, whereas to remelt the same lava it would take a temperature of 1,100 to 1,200 degrees centigrade.

When lava continues to flow under a frozen surface it may drain away, leaving a hollow crust. A number of the pahoehoe tubes which we see at the Craters of the Moon were emptied in this fashion. By the same process large caverns have been produced also. One such cave, the Great Owl Cavern under the surface of the Blue Dragon Flow, is over thirty feet high in places, and two hundred feet long. There are a number of caves of various size at the Craters of the Moon Monument. The little stalactites of lava

CRATERS OF THE MOON NATIONAL MONUMENT, IDAHO. Pahoehoe lava, showing empty lava tubes. *Photo by John Shimer*

found projecting from their ceilings were formed by the hardening mass dripping in freshly drained caves. Many caves collect winter moisture which may last well into the summer in the form of ice; on a hot day a visit to one of these ice caves can be most welcome. Sometimes the roof of a cave will collapse in such a fashion as to produce a natural bridge, and in other places a flow may be dented with collapse depressions due to the falling in of the roofs of lava tunnels. These depressions may range from a few feet up to a hundred or more feet across.

In addition to some very handsome caves, the Blue Dragon Flow possesses a number of peculiar cylindrical hollows, a foot or so in diameter, with a checkerboard pattern mysteriously impressed on their inner walls. These lines were imprinted in the lava when it flowed around a branch or trunk of a tree whose surface had been charred. The wood was later completely burned or rotted away, and now just the hollow remains. One would naturally expect that such an inflammable substance as wood would have burned before it could impress its pattern on the lava, and so we must assume that the lava was relatively cool by the time it overflowed the trees here.

Near the entrance to the Monument there are two other features which merit attention. Lava from the North Crater overflowed at one time and as it poured out it carried parts of the rim with it. The landscape as we see it today shows a black sea of lava with the crater rim remnants emerging from the surface, "bergs" of rim rock on a lava sea.

The Devil's Sewer Pipe is a cylinder of lava about six inches in diameter which can be seen lying neatly along the bottom of a crack, about four feet below the surface. It was formed when a long cylindrical mass of pasty lava was squeezed up into a crack in the frozen surface of the flow. The outside of the lava cylinder solidified while the liquid center drained away, leaving this lava pipe, which here has a visible length of over one hundred feet.

The approach to the Craters of the Moon from the east, starting at Blackfoot on the Snake River, leads across a flat arid lava plain, which extends to the horizon on the south and west and is bounded on the north by a distant range of mountains. Three ages of flows can easily be noted on the drive. Near the eastern margin of the area there is a series of lavas of soft shades of gray, red, and purple. They are the oldest, and their surfaces have now been broken down into soil which will support vegetation, which requires irrigation in this arid area.

The second series of flows which overlie the light-colored lavas is black, very porous, and is now largely covered by sagebrush. It occurs in low billowy wave-shaped ridges cracked open along the top. From a distance these flows look like frozen, seaweed-covered, green waves. The third and last series of flows are those found at the Craters of the Moon. They can

CRATERS OF THE MOON NATIONAL MONUMENT, IDAHO. A recent flow with remnants of rim material carried along as "bergs" in the flow. *Photo by Devereux Butcher*

be seen to overlie the sage-covered lava as we approach the Monument, roughly eighty miles from Blackfoot. These later flows cannot be culti-vated, there is essentially no soil and the surface is extremely rough. Such a jagged surface, which is very difficult to travel over, is called "malpais" (Spanish, bad country) in the Southwest and Mexico.

Both the second and third flows can be seen to finger down the very gentle slopes at their front edges, the fingers in places being hundreds of feet long. Obviously they were composed of a very fluid kind of lava which flowed down the slopes carrying its hard crusted surface on its back. The top surface of the second, now sage-covered, flow wrinkled up into "pressure ridges" as a result of the subsurface drag of the moving liquid underneath. These wave wrinkles, which form a very noticeable part of the scene, are about ten feet high, from twenty to thirty feet wide, and from fifty to a few hundred feet long. The flows of all three ages which now cover most of the region emerged from cracks in the plateau and are unconnected with any visible cone.

As seen from the Craters of the Moon, the margin of the lava flood against the mountains to the north follows a line almost as straight as a contour line, and off to the northeast, a number of miles away, an isolated sage-covered mound of older lava can be seen sticking up as an island in the sea of black froth.

FEATURES such as we have been observing are found in many other volcanic areas of the United States. The various kinds of flows, the caves, the cinder cones with their sprinkling of bombs, the spatter cones, and the tree molds can all be matched elsewhere with local variations.

Wherever volcanic activity takes place it inevitably interferes with the orderly process of erosion. New land surfaces are produced which often totally obscure preexisting surfaces, and former rivers are dammed up to produce lake basins. Such changes have occurred many times in the history of the Columbia Plateau, and they explain the presence of surprisingly extensive lake deposits, especially in the central part of the area. In places the sedimentary deposits have been overlain by later flows, and in other parts the more recent sediments still lie at the surface. The lakes have now been largely drained, but their former presence is indicated by the extensive flat alluvial deposits which are being cut into by present-day stream erosion.

The most recent volcanic events in this area were the building up of the Cascade Range and some isolated activities, such as that at the Craters of the Moon.

The Snake River is now cutting through some of the earlier flows, and in one place at Hell's Canyon on the Idaho-Oregon border it has carved a canyon over a mile deep, through many layers of lava and into the basement rocks, thus rivaling in depth the Grand Canyon of the Colorado River in Arizona.

The sheets of lava which built up the plateau vary in thickness from ten to two hundred feet, and on recently cut canyon walls as many as twenty layers have been counted in a single exposure. Usually, however, not more than a few layers are visible. The lava is so widespread and uniformly distributed it must have come up through innumerable cracks and crevices. Probably none of the feeder cracks was very wide; furthermore, they were most likely quite uniform to a great depth. Such a conclusion was reached in a study of similar feeder cracks on the Hawaiian Islands, where they have been exposed to a depth of over 4,000 feet in places; to that depth at least they show no appreciable variation in their width of about five feet.

Considered as a whole the Columbia Plateau is one thick mass of lava, made up of many flows, one on top of another. However, as we have seen, the picture is complicated in places by the presence of sedimentary material, and by the embellishment of the land's surface by a variety of recently produced features such as those found at the Craters of the Moon. In its details the picture has been further confused by the fact that weathering and erosion have been continuously active ever since the first lava appeared, and diastrophism has tilted and crumpled some of the earlier flows, especially in the western part of the plateau.

MOUNT ST. HELENS, A VOLCANO IN THE CASCADES OF WASHINGTON. This is a strato-volcano. *Courtesy Department of Commerce and Economic Development, Olympia, Washington*

The Columbia River east of the Cascades has cut well into the volcanic flows, which are mostly composed of black nonporous lava, basalt. The scenic drive down the Columbia as it follows the state boundary between Washington and Oregon furnishes marvelous views of these flows. Seen edge on, as exposed on the walls cut by the river, they often appear to be composed of vertical stubby columns which look almost like a man-made wall in their perfection.

Such columnar jointing is found in sheets of basalt whether they emerged as flows or were forced in as a liquid between preexisting layers of sedimentary rock to form what are called "sills." This kind of jointing accounts for some of the most exotic types of scenery found in the world. The columns may be remarkably regular, are often hexagonal, and may be a few inches to a number of feet in diameter, depending on the thickness of the flow. On cooling, a sheet of lava can contract in the vertical direction very easily, but contraction in the horizontal causes cracks in a way somewhat like that in which mud cracks are formed. The Devil's Postpile in California, the Giant's Causeway in Northern Ireland, the fluted appearance of Devil's Tower in Wyoming, and the Palisades sill on the west side of the Hudson River at New York City are all notable because of their columnar jointing.

At Bend, Oregon, we are at the western extremity of the Columbia Plateau; before us stretches the majestic mountain mass of the Cascades. An excellent view of this range and the surrounding countryside can be obtained from the summit of Pilot Butte, a small cinder cone east of the town. All the visible landscape was predominantly molded by volcanic forces now dormant. To the west the Oregon Cascades form a jagged and snow-covered rampart. Mount Hood with its beautiful classic shape is visible 100 miles away to the north, and beyond it, over the horizon in Washington, we can only imagine such peaks as St. Helens, Rainier, and Baker. South of Mount Hood, and nearer to us in Oregon are Mount Jefferson, Three Fingered Jack, and Bachelor Butte, the last only about twenty miles away. All of these mountains are a part of the volcanic chain which ends in northern California with Mounts Shasta and Lassen. In the immediate foreground to the south there is a whole rash of cinder cones littering the countryside. They are covered and somewhat masked with an extensive coating of evergreen trees. Dozens of cones are in view, each one marking a channel through which hot liquid and gases were spewed from the depths of the earth. Immediately surrounding our lookout is a flat volcanic plain, not very wide toward the Cascades and the cinder cone area, but broadening toward the north and east until mountain islands block the view. The lava goes around them, and we can think of the sea of lava as essentially extending to the Grand Tetons of Wyoming over 500 miles away.

The cinder cones which we see are a part of a group of over 150 which have grown up on the flanks of a very broad, inconspicuous, low lava dome, called Mount Newberry. This large dome, twenty miles in diameter at its base and 4,000 feet high, was constructed almost entirely of fairly liquid lava with very little solid debris. The cinder cones here, which are parasitic growths on the older shield-shaped Mount Newberry, were formed during the terminal stages of vulcanism when there was a more explosive type of activity and solid instead of liquid material emerged from the ground.

Mount Newberry resembles very closely the volcanoes of the Hawaiian Islands, which also were almost entirely built up of liquid material and have a typical low convex shield shape. At the summit of Mount Newberry there is a large depression 5 miles across in its largest dimension, with cliffs on one side 1,500 feet high. Two small lakes are now located in this hollow. Such a depression at the top of a volcano is called a "caldera" (Spanish, cauldron). This caldera was caused by the enlargement of the original much smaller crater when its walls collapsed into the throat of the volcano as the liquid was partially withdrawn at the end of active vulcanism. The Hawaiian volcanoes also have such calderas at their summits. That of Mauna Loa on the Island of Hawaii is two miles wide and about

1,000 feet deep. Kilauea, also on Hawaii, has a caldera one and a half miles wide, in a part of which a liquid pool of lava is almost constantly bubbling. This fire pit of Halemaumau is one of the most notable features of this volcano.

The shape of the average Cascade volcano is strikingly different from the low rounded appearance of a shield volcano. The Cascades were built up of alternating layers of liquid and solid, flows and cinders, and are thus intermediate between the shield volcano, with its primarily liquid composition, and the small cinder cone, with its essentially all-solid debris. Both the stratified cones as well as the shield cones may in places be attended by parasitic cinder cones. And in both, many of the flows, instead of coming out of the top of the volcano, emerge from the sides or base of the cone when the main passageway is plugged by solidified lava.

The parasitic cinder cones around Bend, Oregon, are frozen relics from a time of dynamic action. Similar action in a very similar scene occurred recently in Mexico when the large cinder cone of Paricutin was born, lived for about ten years, and then died. A description of the happenings at Paricutin should help in mentally reactivating the Oregon scene with growing cinder cones.

Paricutin is a fairly large cinder cone in a region of hundreds of such cones. Its physical growth started on February 20, 1943, at about 4:00 PM when an Indian farmer, Dionisio Pulido, noted wisps of "warm smoke" coming up through cracks in the ground. This was followed very soon by pieces of rock and cinders, very much as if someone were down in a hole tossing such material at random out into the air. The cinders came out at such a rate that by the next morning a cone 100 feet high was produced. In two weeks it was 500 feet high and in a year it had reached essentially its maximum height of 1,400 feet. The material which emerged at first was

PARICUTIN VOLCANO,
MEXICO. *Courtesy*
American Museum
of Natural History

primarily in a solid state, as cinders, dust, and volcanic bombs. At night the bombs glowed like rockets as they arched up into the air to fall on the slopes of the cone and come rolling and bouncing down leaving a trail of sparks behind. Later on lava broke through the base of the cone and inundated large tracts of the surrounding land. The flows were primarily of the aa type of blocky angular lava. One small village, that of Paricutin, about one and a half miles away, was entirely covered by lava. About half of the town of San Juan, over three miles away, was also eventually submerged. It was here that the flow finally came to a stop a very short distance beyond a large stone church which now lies engulfed in the irregular lava blocks above which the twin towers and some of the walls still protrude.

For about two weeks before the actual start of the cone's growth the ground of the area was shaken by many earthquakes, and on the day before there were about 300 such quakes.

At the same time that the cone was growing and the flows were spreading locally over the countryside, a great deal of fine black dust was emitted and was spread broadcast over the surrounding area. This very fine ash and dust was more than ten feet deep two miles from the vent, and was over one half foot deep ten miles away. It covered and suffocated most of the vegetation so that now there are large tracts of desert area covered with nothing but black dust and ash. Where this very light material has been removed by rainwash from the slopes, the native evergreen type of vegetation has been able to reassert itself.

The last lava was extruded in 1953. Three years later the region was almost quiet again. However, a little steam continued to emerge at the surface from the depth of one of the flows. Such jets of gaseous material at the surface very often look like smoke emerging from the ground. It is not surprising, therefore, that such features are called fumaroles (which derives from the Latin word, *fumus*, smoke). Fumaroles are obviously associated with volcanic activity and are composed primarily of superheated steam, with sulfur compounds, hydrogen gas, hydrochloric acid, carbon dioxide, ammonium chloride, and a host of other materials. During wet seasons, when there is a great deal of dilution by rainwater, percolating down into the ground, a fumarole may be essentially a hot spring, with some steam and some water. But in dry seasons the hot spring may stop flowing and superheated steam, perhaps coming directly from magma, emerges at the surface.

The "smoke" which heralded the birth of Paricutin, and which some-times can be seen to leave the throat of an active volcano, is also usually composed of condensing steam which may also carry fine rock particles in suspension.

With the story of Paricutin fresh in our minds it is comparatively easy to visualize the landscape in the region of Bend in the days when the cin-der cones were growing. Probably not more than a few were active at any one time, as such cones are generally very short-lived, becoming extinct within a few years.

SOUTH of Bend, we enter a region where vulcanism has created yet an-other kind of scenery. Crater Lake, in Oregon, is a truly remarkable sight. It is located in the enlarged crater of an ancient volcano. In order to reach the Lake, one must drive uphill for miles before attaining the rim, which is over 7,000 feet above sea level. Unlike most lakes, it is not possible to see this one until we are actually on the surrounding rim. It is six miles across, lies 2,000 feet below some of the higher parts of the jagged rim and is about 2,000 feet deep. The color of the water as we look from a vantage point far above its surface is a unique and wonderful deep blue.

The first question naturally that one asks is, how was the lake basin formed, how was such a large and deep crater or caldera produced? Other questions arise after a closer inspection of the scene, especially after a lei-surely and enquiring drive around the lake.

From the rim on the southwest side in the vicinity of Rim Village, we can obtain an excellent view of a number of most interesting features. Toward the north, in the western end of the lake lies Wizard Island, a beautiful little cinder cone with its summit about 800 feet above the level of the water. We look down on top of the island from the rim over a stretch of the blue water. The cliff face of the caldera beyond the island shows a layered structure, the layers dipping into the cliff face—that is, away from the lake. It is apparent that these layers at one time must have extended much farther in over the area now occupied by the lake. In places the layers of alternating lava and ash are obscured by piles of talus and slide rock which slope down smoothly into the water.

The massive cliff of Llao Rock appears just to the right of Wizard Island.

CRATER LAKE, OREGON, fills the enlarged crater of an extinct volcano. A small cinder cone forms Wizard Island. To the right of this island is the steep cliff of Llao Rock, and to the left the Devil's Backbone, which is the exposed end of a resistant dike.

Courtesy Oregon State Highway Commission

The cliff has an odd shape. It is a wedge of massive lava which has a rounded U-shaped base and projecting tapering parts on either side. If Llao Rock itself were removed the rim here would show against the sky as a U-shaped depression, which is actually the end of a U-shaped valley which continues down the mountainside and is now filled to overflowing with lava. Such a valley should of course remind us strongly of a glaciated trough.

To the left of Wizard Island a narrow ridge of rock extends from the top of the cliff to the water's edge. This is the Devil's Backbone, a lava dike. It stands out in conspicuous contrast to the softer flows and cinders in which it is found, and into which it has obviously been intruded.

The drive around the lake rim is over thirty miles. From many places on the drive beautiful views can be obtained over the lake to the opposite rim, which seems to look different from each vantage point and at different times of the day. From a few places it is possible to get a view out over the surrounding countryside and to get a feeling of the size and shape of this large volcano in which Crater Lake is hidden.

One of the interesting sights from the drive is Phantom Ship, a small island near the rim on the south side of the lake. The backbone of this islet, the "sails" of the ship, is composed of resistant dike material, which, like the Devil's Backbone, is part of one of about a dozen dikes which cut the wall of the crater.

On either side of the Phantom Ship the rim can be seen to have a U-shaped sag, and from the eastern side of the lake the Park headquarters area can also be seen to lie in such a U-shaped notch. Each of these notches is found to continue downward as a typical U-shaped glacial valley.

There are thus four obvious glacial valleys cutting the slope of the volcano, one of which has been filled with a lava flow and now shows up as Llao Rock. Glacially striated boulders which have been found in the bottom of this flow are added evidence that glaciers at one time streamed down the slopes of the volcano. The abrupt start of these U-shaped valleys as full-fledged affairs right at the present-day rim and the absence of cirques are features which need further explanation. It seems obvious that the upper parts of these valleys must have disappeared, having lain at one time farther up a mountain which is now represented by the empty space over Crater Lake. We need, moreover, this vanished mountaintop to explain the dikes which now stand out as widely spaced vertical ribs on the rim face, since they must have been enclosed at one time in other lava which they intruded.

The assumption that this volcano was at one time higher, with a smaller-sized opening at its summit, thus rests on such evidence as the present abrupt termination of the lava flows as seen on the rim face, flows which on their continuation must have arched up and out over the present lake, on the sudden end of the U-shaped valleys, and on the bare dikes obviously showing now only because of the retreat of the rim cliff.

These features of the crater are clues to the geological history of the region and tell us of a drama that took place over a million years ago. It was then that Mount Mazama, the ancestor of the Crater Lake volcano, first made its appearance. During a time measured in hundreds of thousands of years, volcanic activity from deep within the earth pushed up Mount Mazama and the other peaks of the Cascades, such as Shasta, Hood, Rainier. Their growth was marked by alternations of active explosions and relative quiet. Sometimes the volcanic forces under Mount Mazama stayed dormant for centuries, when soil and vegetation had time to develop on its slopes. The principal growth of this mountain and its sister peaks of the Cascades occurred in the Pleistocene epoch. Mount Mazama with its height of over 12,000 feet rivaled at that time any other volcano in the whole of the Cascade Range.

It was a typical composite cone built of alternating layers of lava and ash. In the later stages of its growth a great deal of lava welled out near the base and the previous material of the cone became criss-crossed with many dikes as the lava oozed up into cracks and there froze into more or less vertical sheet-like bodies.

Throughout its growth Mount Mazama probably had glaciers streaming down its sides many times and must often have been practically smothered in ice. Such periods were interrupted by volcanic activity as shown by the alternation of glacial and volcanic deposits. Some of the earliest layers of glacial material outcrop near the present lake level, thus representing a time of glaciation before the volcano came anywhere near reaching its maximum height.

After reaching this height and while there were still glaciers streaming down a number of the valleys, the volcano lost its head! This dramatic episode happened so recently, geologically, that the succession of events that led up to it can be reconstructed in detail. For several weeks there were very violent eruptions which spread tremendous amounts of pumice and dust over the surrounding countryside. A foot of such material has been found over seventy miles away; about three miles north of the rim of the lake it is piled to a depth of 250 feet, creating a pumice desert, though 72 inches of water in the form of rain and snow falls annually on this area. Nothing will grow because the moisture just sinks into this very porous material, leaving behind a dry inhospitable land.

The second stage of the self-decapitation of Mount Mazama consisted of a still more violent explosion of incandescent pumice. This was the stage of the "glowing avalanches," when hot waves of pumice literally flowed down the valleys with a hurricane speed of 60 to 100 miles per hour. The pumice fragments rode along on a cushion of hot escaping gases, each fragment surrounded by a self-formed series of spouting steam jets. The hot blast of the compressed air that preceded the flow had tremendous power. It mowed down large trees, whose charred remains can now be found buried in the lower layers of the flow. This flow of pumice covered an incredible distance. A piece fourteen feet long was found twenty miles away from the top of the mountain. Pumice is a very light frothy kind of rock, which will float on water, so that it is not too surprising that such large pieces were moved for such distances. As the particles were swept along, they banged into each other and produced a great deal of fine powder which was incorporated with the larger pieces when they finally came to rest. After this tremendous explosion the many valleys radiating from the crater were each filled with glowing, steaming pumice. An outcrop of this material is now found in the southeastern part of the Reservation on Sand Creek, where it has been eroded into steep pinnacle remnants, some over 200 feet high. Chunks of pumice of various size are conspicuous in the general mass of material which is fairly well consolidated. Initially the glowing fragments welded themselves together when they came to rest, and following this the hot gases which rose from the steaming pile altered and still further solidified vertical pipe-like masses of pumice.

In a very short time after the glowing avalanches, the final event took place. This was when the top of the mountain simply caved into the empty hole formed partly by the blowing out of so much pumice material, but primarily by the withdrawal of a large mass of magma back down into the earth. It has been estimated that approximately ten cubic miles of the top

THE PINNACLES, CRATER LAKE NATIONAL PARK, OREGON, are carved from a thick deposit of pumice fragments. *Courtesy Oregon State Highway Commission*

of Mount Mazama disappeared, of which perhaps only about one and a half cubic miles were represented by the shattered rock and pumice of the last explosions, the remainder disappearing into the earth.

The destruction of Mount Mazama occurred about 6,500 years ago, after the last maximum advance of the glaciers in the Ice Age. The desolation at that time was complete. Vegetation was utterly destroyed on the slopes, glowing avalanches were smoking with great activity, and, where the top of the mountain had been, there was a hissing, vapor-filled cavity of tremendous size, floored with a mass of jumbled rocks choking the throat of the old volcano, which had already gulped down so much lava. This then was the way that the great caldera of the present Crater Lake was formed.

More recently a minor amount of volcanic activity resulted in Wizard Island, rain and snow have filled the depression, and weathering and the force of gravity have modified the rim slightly to produce the talus slopes and reveal such features as Phantom Ship and the Devil's Backbone.

Since it is known that people have lived in our Pacific Northwest for over 10,000 years, it is probable that human eyes witnessed this tremendous drama. Perhaps what they saw will seem more real to us if we compare it with some historic volcanic explosions which were very similar to what undoubtedly happened here in connection with Mount Mazama.

On May 8, 1902, Mont Pelée on the island of Martinique in the West Indies exploded and shot a large mass of dust, pumice, and red-hot rock far up into the air to form a reddish, black, seething cloud. When it settled and hit the ground it shot down the slopes of the volcano in a glowing avalanche, burning and destroying and suffocating everything in its path. The French colonial city of St. Pierre, with its fine buildings and parks and its 28,000 inhabitants, was utterly destroyed. The hot hurricane blast of air which preceded the flow of pumice laid flat everything in its path, and the hot pumice set fire to everything burnable, and the escaping toxic gases suffocated all living beings left. It was one of the most sudden and devastating disasters ever recorded. The glowing cloud of material was more than 650 degrees centigrade, because melted glass vessels were found. Eyewitness accounts by those outside the town and from the various ships in the harbor referred to the wall of fire which swept down the slope, and to the tremendous noise, like that of hundreds of cannon. There were 18 ships in the harbor at the time of the eruption, and of these 16 were destroyed first by the hot blast and then by the fire.

It was a blast similar to this which probably preceded the collapse of Mount Mazama. After the glowing avalanches had come to rest on the slopes of the mountain, they steamed with thousands of noxious fires, and each valley undoubtedly resembled the Valley of Ten Thousand Smokes in Alaska, which came into being after the pumice eruption of Mount Katmai in 1912. This eruption was a complete surprise as Katmai had been dormant for so long that it was considered virtually dead. Fortunately it was in a very sparsely settled area along the Alaskan coast, and, as far as it is known, there were no lives lost. The explosion shot into the air a great deal of material which traveled for very many miles. Over one foot of ash fell on the town of Kodiak 100 miles from the volcano. The sea nearby was covered with floating pieces of pumice, many square miles being covered with a layer thick enough to support a man.

One of the most interesting features associated with this eruption, however, was the production of a tremendous field of steaming pumice sand. Literally several million fumaroles were busily smoking away in an area of about 50 square miles. This was the Valley of Ten Thousand Smokes. Long after the pumice settled in this valley it continued to emit enormous quantities of gas. Six million gallons of steam were produced each second, and each year, 1,250,000 tons of hydrochloric gas and 200,000 tons of hydrofluoric acid were emitted. This gives some idea of what the slopes of Mount Mazama must have been like after the glowing avalanches had come to rest.

THERE ARE probably somewhat less than 600 active volcanoes in the world at the present time, that is, volcanoes which are currently functioning or dormant with a strong likelihood that they will be active again in the near future. Most of the active cones are located in well-defined zones on the earth, and undoubtedly follow lines of weakness in the crust.

The volcanic area of the western United States which includes the Cascades is part of the Ring of Fire, a belt of igneous activity which surrounds the Pacific Ocean. This is the world's major volcanic zone and stretches from Cape Horn up through the volcanoes of the Andes, the volcanic area of Mexico, the Cascades, the Alaskan volcanoes, which include Katmai, the Japanese islands, and so down through the Philippines to New Caledonia and New Zealand. Other shorter stretches of volcanic activity are found in the Mediterranean belt which extends from the Azores to Mount Ararat in Asia Minor through the Italian volcanic zone; the Rift Valley area of Africa; the East Indies area of Timor, Java, and Sumatra; and various isolated patches such as Iceland, the Hawaiian Islands, parts of the West Indies. This present-day distribution of volcanic activity is also roughly the distribution of the major earthquake belts and the areas of some of the large mountains. Diastrophism and vulcanism thus often appear together as manifestations of crustal unrest.

In the United States, recently formed scenic features due to vulcanism are found exclusively in the western part of the country. They are concentrated, as we have seen, in the Cascades and the Columbia Plateau, but also occur with notable examples in the Colorado Plateau, Utah, Nevada, Wyoming, and California. They may also be found associated with older intruded igneous rock forms whose shapes have been exposed by erosion. Such exposed skeletons of intrusive rocks can control the development of scenic forms for a very long time.

The volcanic modifications of the Colorado Plateau illustrate admirably both the extrusive and the intrusive igneous rock forms. South of the Grand Canyon, near Flagstaff, the flat surface of the Plateau is interrupted by a series of volcanic cones, and chunks of scoria which resemble giant-size pieces of coarse black bread are scattered widely over the land. Some of the cones are so recent that the crater at the top is still largely intact. Sunset Crater is an especially fine example, and has been set aside as a National Monument. It is composed of dark red cinders near the top with even darker colored ones near the bottom. This cone is believed to be less than 1,000 years old. It is about 1,100 feet high, a quarter of a mile across, and 400 feet deep. A lava flow showing a coarse crusted surface emerged from the base. There are some spatter cones above the flow and a few caves in it.

Ship Rock in the northwest corner of New Mexico stands out from a dead flat surface. It is a jagged pinnacle from which radiate a number of uneven narrow ridges composed of igneous material. When viewed from a distance and at certain angles this rock assemblage looks very much like a ship sailing across a quiet sea. In this case erosion has entirely removed a former volcano and cut the surface of the land down to expose the lava frozen in the throat, as well as two clearly defined resistant ridges radiating from the central peak. These ridges are the frozen remnants of lava squeezed out of the central conduit into the surrounding rocks which were cracked open by the pressure of the ascending lava; thus they are similar in origin to the dikes of the Crater Lake rim.

Many of the National Parks and Monuments of the West have notable volcanic features, which essentially are only variants on the basic types of volcanic scenery discussed so far. Columnar jointing, cones in various stages of removal, black cinders of scoria, and white cinders of pumice are some of the most obvious scenic elements to be encountered.

The Lassen Volcanic Park of northern California is rich in cones and flows. Lava Beds National Monument, also in northern California, has many lava caves, natural bridges, cinder, and spatter cones. And other cones and flows may be seen widely spaced in the Basin and Range Province of Nevada and Utah.

Yellowstone Park has some spectacular volcanic scenery. The Geysers and hot springs are discussed in a later chapter, but we should not pass through the Park without stopping to look at Obsidian Cliff. Here there is a huge mass of shiny black glass which forms a notable feature along the roadside. A lava with no dissolved gases in it emerged from the earth here and chilled suddenly to become obsidian. We can pick up small fragments of it, sharp-edged and glistening, and can easily see why the Indians considered this the finest material for arrowheads.

In another part of Yellowstone Park, near Tower Junction, we can see some especially fine columnar jointing. When we look at the straight, even pillars of rock standing side by side, cutting in a level line across the cliff face, they seem more like part of a man-made wall than anything else, and it scarcely seems possible that these symmetrical columns are just the result of the cooling of a lava flow.

Columnar jointing even more astonishing than that at Yellowstone is to be found at Devil's Tower, Wyoming. Here a mass of igneous material

800 feet in diameter soars into the sky to a height of 600 feet above a broad talus slope of broken debris. It represents a denuded mass of frozen magma which was pushed into a series of horizontal sediments. On cooling, it contracted and produced the typical columnar jointing. In this case, however, the joints are especially large and impressive and give the tower from a distance a vertical fluted appearance. There has been some difference of opinion as to the actual origin of the Tower. Some geologists feel that it is the exposed neck of a former volcano, similar to Ship Rock in origin. Other geologists consider that its shape has always been

DEVIL'S TOWER, WYOMING. This fluted mass of volcanic rock towers 865 feet from its apparent base. *National Park Service Photo*

about the same as it is now and that it merely represents a bit of cooled magma pushed up like a finger into overlying sediments, now removed by erosion.

The actual development of a mass of liquid rock, or magma, in the depths of the earth is probably due to a combination of two factors, relief of pressure and radioactive heating. There is a rise in temperature with increasing depth everywhere in the earth. It does not take a depth of many miles before a temperature is reached which is hot enough to form a magma. However, the melting temperature of rocks is raised by pressure, so that material at a high enough temperature to be liquid at the earth's surface will still be solid at depth. This potentially liquid material will change into magma, however, when the pressure is released, say, because of faulting, and may then eat its way upward to the surface where it will emerge to form a volcano or flow.

There is an extra source of heat in the form of various radioactive elements which may be locally concentrated in the crust and which even under pressure may by their spontaneous decay provide heat enough to melt rocks. The resulting small patches of magma are purely local accumulations in the otherwise solid outer parts of the earth and are totally unconnected with the liquid at the core of the earth.

There are two kinds of magma; one produces the light-colored igneous rocks, the light gray, and sometimes reds and greens, and the other the dark gray to black rocks. The difference in color is due primarily to a difference in chemical composition. When magma appears at the surface of the earth it is called lava. If a flow does not freeze immediately it will crystallize into a fine-grained rock, in which there are mineral grains perhaps too small to be seen with the naked eye but perfectly obvious under a microscope. Basalt and the lighter-colored rhyolite and andesite are such rocks. A glassy texture, on the other hand, means very rapid cooling at the surface of the earth, and is produced only when a magma is extruded as lava flows or is shot out from the throat of a volcano as globules of liquid and solidified immediately. If there is a great deal of gas dissolved in the magma, glass froth, such as the black scoria and the lighter-colored pumice, is formed. If, on the other hand, there is no gas, dense glass-like obsidian is produced.

Basalt flows are quite liquid and spread out into flat sheets, such as those that built up the Columbia Plateau, or they may form such low shield-shaped domed volcanoes as Mount Newberry in Oregon. The more viscous, light-colored lava freezes on a much steeper slope and sometimes, acting like very thick toothpaste, bulges out into very odd shapes. Following its catastrophic explosion Mont Pelée produced a very unusual spine-like projection which was pushed up out of the throat of the volcano. It was composed of this very thick lava and was extruded at the average rate of 40 feet a day until a thousand-foot pinnacle stuck up into the sky. The pasty liquid froze almost as quickly as it emerged and thus maintained its shape. It was a very unstable edifice and was soon broken.

Similar extrusions, which have taken a collapsed mushroom shape, form bulbous plugs in a number of the volcanoes of the West. The Mono Craters of eastern California, just at the eastern foot of the Sierra Nevada Mountains, have such plug domes of viscous lava, whose shapes can be easily discerned at a distance. These cones lies along a slightly curved line, which undoubtedly traces a major crack or fault in the crust. A similar lineup of the spatter cones in the Craters of the Moon area has been noted. In fact it is very common for a series of volcanoes to show such an arrangement, along a major break in the crust which taps a magma source.

We have seen now that the most exciting aspects of volcanic scenery occur in the West. We cannot find their exact counterparts in the East. Vulcanism in the eastern part of the United States took place long before the earliest stages of the growth of the Columbia Plateau—so long ago that any surface volcanic rocks formed at that time have been either removed or buried under sediments. Now, however, after extensive erosion some of these buried rocks have been uncovered. The Watchung Mountains of New Jersey and some of the ridges in the Connecticut Valley in Massachusetts and Connecticut are the uptilted eroded edges of buried lava flows. Obviously these ridges did not exist until after erosion had removed the softer material from around them. The Palisades Ridge west of New York City in New Jersey is a tilted mass of lava also, but in this case instead of being a flow it was squeezed while still liquid between previously existing layers of sediments into a flat tabular body called a sill. Following tilting and erosion, both sills and flows will form ridges, very similar in appearance.

Dikes, thus far mentioned only in connection with volcanoes, are very common in most deeply eroded mountain areas also. They are produced when magma oozes into a crack below the earth's surface and there freezes, forming a fin-shaped igneous rock body with relatively straight walls. Some dikes have been found to be only a fraction of an inch across. However, there are a number which are hundreds of feet wide, and may be exposed in places for a number of miles.

The major bulk of any magma never reaches the surface but cools at depth into igneous rock bodies of various shape, which we can investigate only after weathering and erosion have removed the rocks which overlay the cooling magma. As we have seen the presence of a dike, a sill, or granite at the surface of the earth indicates such erosion.

When we see a granite ledge we are looking at a rock which solidified under perhaps thousands of feet of overlying rock layers.

Granite appears on the surface in the deeply eroded parts of New England. It forms the older, higher parts of the Appalachian Mountains, the Rockies, the Sierra Nevadas, and is exposed at the bottom of the Grand Canyon in Arizona. It weathers in a number of different ways, and thus is responsible for various types of scenery. As we have seen in the Sierra Ne-

DIKE OF BASALT CUTTING GRANITE, COHASSET, MASSACHUSETTS. *Photo by John Shimer*

vadas at Yosemite Park, it has weathered to form the characteristic massive cliffs and the exfoliation domes. In the Central Black Hills in South Dakota, the granite core weathers in a quite different manner. Here because of a series of closely spaced vertical joints the rock has been worn into vertically standing pillars or needles. Most usually, however, granite will weather into rough irregular shapes. Jointing will generally be obvious and run in a number of directions and show a definite control over the angular rough topography which erosion produces.

Another dike-shaped body which is commonly associated with granites is the pegmatite. A granite pegmatite has a very different origin from the ordinary dike and also differs in having rather irregular walls. It may vary markedly and abruptly in thickness from one end to another and is composed of crystals of greatly varying size, from a fraction of an inch to perhaps many feet across.

The Black Canyon of the Gunnison in central Colorado is notable for its depth and narrowness and for its Painted Wall, where the north wall of the canyon is crisscrossed with pegmatite dikes. From one place on the south rim it looks as though some artist had spilled his paints and the light shades had all dribbled down the cliff, as the white to pink pegmatite streaks stand out very dramatically from the surrounding dark gray rock. Here we just see the exposed ends of the pegmatite dikes which we should imagine as crisscrossing through the whole mass of the granite.

Ordinary igneous rocks formed by the solidification of a magma do not have such large crystals as those found in pegmatites. Magma when it is forced up to form a dike always cools fairly quickly, intruded as it is into cold solid rock. The minerals in a pegmatite, on the other hand, grow slowly from material carried in very hot aqueous solutions, and are commonly composed of quartz, feldspar, and mica. Such pegmatites are found in all kinds of granites and sometimes extend up into the rock which the granite intrudes.

IT must be apparent by now that an important part of understanding scenery consists in appreciating the structure of various underground rock masses. Thus the Painted Wall should be thought of not as a surface feature but as a series of three-dimensional bodies seen edge-on. The shape of such dikes may often resemble the shapes seen on a slice of marble cake.

At first it was thought that igneous rocks were part of the initial crust of the earth, formed when our planet's skin first developed, after a presumed gaseous and then liquid phase of its history. However, if there was such an initial crust, it has not been found. Wherever igneous rocks occur they can be seen to cut still older rocks, sedimentary or metamorphic. The origin of the earth is not known, and its early history is hard to decipher; only the recent events are really clear. As far as can be determined, however, igneous activity has been a force to be reckoned with throughout all of known earth history, first active in one place in the crust and then in another, and it is obviously responsible for much of the more striking scenery of the present world.

THE GRAND TETONS, WYOMING   *Union Pacific Railroad Photo*

CHAPTER 7

MOUNTAIN landscapes are largely the product of erosion. Driving in a region of peaks, slopes, cliffs, and gorges, we become increasingly aware that the shape of the land is new. We see the results of recent landslides, and frequent signs warn us against fallen rocks in the road. Swift-flowing streams, bare rock ledges, and steep slopes are all indications of the rapid rate of destruction of the land.

Mountain Scenery

As landscape features go, mountains are certainly young. All have been recently uplifted and are now being rapidly removed by erosion. This rapid removal is so evident that it is obvious that they have been uplifted in the very recent geologic past by the counterforces of diastrophism and igneous activity. In the mountains where the naked rock bones of the earth have been washed and scraped bare of any soil covering the results of the awesome force of diastrophism are especially obvious. We can see crumpled and contorted rock layers, joints which crisscross every kind of rock, and in some places faulted structures. The product of igneous activity of long ago may be noted in the exposed edges of dikes and sills and in the outcrop of large masses of granite. And elsewhere more recent igneous activity may have produced a cone or lava flow.

The two major areas of mountains in the United States lie in the eastern and western thirds of the country. The mountainous tract in the east is far

FOLDS IN LIMESTONE NEAR PLACER DE GUADALOUPE, CHIHUAHUA, MEXICO. Note the syncline at the left and the anticline at the right. *Photo by Kirtley Mather*

JOINTING IN GRANITE, ON ROUTE 128 NEAR DEDHAM, MASSACHUSETTS. The joints form planes of weakness along which the rock broke when it was blasted for a new road. *Photo by John Shimer*

less extensive and has a more subdued aspect that its western counterpart. It consists primarily of the two sections of the Appalachians, the high and resistant Blue Ridge and the Great Smoky area on the east and the Folded Ridge and Valley part on the west. The Blue Ridge is separated from the coastal plain by a piedmont area and the Folded Appalachians are bordered on the west by the Appalachian Plateau, an area of moderately deep valleys and narrow ridges. Other small isolated mountain patches of the East occur as the Adirondacks in upper New York State, the Green Mountains of Vermont, and the White Mountains of New Hampshire, which were discussed in connection with glaciation.

The Blue Ridge-Great Smoky area and its piedmont region is composed of a mass of ancient metamorphic and igneous rocks, which are all that is left of a chain of long vanished mountains. They now survive above the coastal plain because of their great resistance to erosion.

Just to the west of this ancient area, younger sedimentary rocks have been crumpled into a sequence of folds.

This belt of wrinkled rocks and their adjoining plateau area extends over a thousand miles from New York State to northern Alabama. In Pennsylvania the range of folded mountains is about eighty miles wide, farther south from thirty to forty miles, and to the north in the Hudson Valley Lowland it is only a couple of miles wide.

From Harrisburg to Pittsburgh the Pennsylvania Turnpike crosses both the folded part of the Appalachians as well as part of the plateau area toward the west. Along the Turnpike we can find excellent views of the various rock layers, and appreciate the relationship of scenery to the rocks and their structure.

On the section of approximately eighty miles between Blue Mountain and Allegheny Mountain we cross a number of sharp crested ridges, with intervening deep valleys. The road goes through a number of the ridges by tunnel, and at other places through deep cuts made in the sandstones and conglomerates of which the ridges are composed. We see these rocks in layers which dip first one way and then the other. It seems incredible to us that these tilted layers are actually parts of gigantic folds; that these sedimentary rocks, formerly laid down in some shallow sea that washed against ancient mountains to the east, have been crumpled and squeezed together by the enormous action of earth forces; and that now only the truncated base of the folds is visible.

The Appalachian ridges furnish an impressive example of the power of diastrophism. In one place, for instance, sediments that formerly stretched in horizontal layers for eighty-one miles have been compressed to sixty-six miles after folding. Altogether there are about eleven major folds in this part of Pennsylvania. The sharp-crested, steep-sided ridges, called "hogbacks," are residual remnants of a relatively hard layer of sedimentary rock which occurred between layers of less resistance. Such ridges are produced wherever sedimentary rocks are tilted or folded. All the Appalachian ridges have approximately the same elevation. If we leave the turnpike and go to the top of one of them, we can readily note the very even line which the neighboring ridges make against the skyline. The valleys which we cross between the ridges are cut into shales or limestones. These softer sedimentary layers have been worn away by erosion, leaving the more resistant layers behind as ridges.

If we are very observant we will note that here, in this central part of Pennsylvania, geology has in large measure determined the location of the highways. Many of the roads take advantage of the valleys and thus show a consistent pattern with a northeast-southwest trend. The northwestern part of the state is in the plateau area of the Appalachians, and the road patterns and valleys, reflecting this different structure, are far less regular in their orientation.

On a map one can also note the very remarkable paths of some of the major rivers, such as the Susquehanna and the Delaware, as they cut straight across the folded structure and form a narrow water gap each time they cut through one of the resistant ridges.

The Delaware Water Gap is the most famous of these. It occurs where the Delaware River cuts through one of the most resistant hogback ridges of the whole area, contradicting the general rule that rivers cut their valleys where the rocks are weakest. For an understanding of this phenomenon we must review the events that led up to the present aspect of this Appalachian region, beginning about 500 million years ago.

The history of any typical mountain range consists of three parts: rock-making, mountain-making or diastrophism, and finally, mountain-carving. The rock-making episode of the Appalachians was continuous throughout most of the Paleozoic Era, roughly for 300 million years. During this time much of eastern North America was periodically flooded by shallow seas in which was dumped a thick sequence of rock debris, gravel, sand, and mud, and in which, farther from shore, lime was also precipitated. In places over 40,000 feet of sediments were deposited as the area slowly sank. Toward the end of the Paleozoic when this part of the earth for brief times was above the ocean, thick forests grew on the low-lying flood plains and swamp areas. Each time a rise in sea level caused the burial of these forests by sediments, a future coal seam was formed. The subsidence of the area was spasmodic, which explains the presence of terrestrial or swamp and flood-plain deposits interbedded with marine deposits. Probably the coal swamps of that age resembled in many respects the Great Dismal Swamp of Virginia and North Carolina of today, where 1,500 square miles of land near sea level have a thick vegetation growth.

Such a tremendous area on the earth which sinks throughout a long time is called a "geosyncline."

The Appalachian geosyncline was roughly 1,500 miles long. It extended from Canada to the Gulf of Mexico, and had a width of 300 to 400 miles. Most of the material that filled it was washed down from former highlands to the east, roughly where the present Coastal Plain is.

As the sediments piled up deeper and deeper, the earlier layers were compressed by the weight of the overlying load and during the long course of geologic ages became hardened into rock. In the case of mud, water was squeezed out of the initially semiliquid material and the microscopic flaky particles were compressed to form shale. The original layers of gravel and sand were turned into conglomerates and sandstones by pressure aided by the process of cementation by natural cements, such as silica and the various iron compounds.

The rocks of the Appalachians were thus produced and waiting to be molded into a mountain range by the end of the Paleozoic Era, somewhat over 200 million years ago. At that time a tremendous change overtook the region and the mountain-making stage was initiated. Instead of being periodically flooded by the sea, as it had been for so long, this part of North America was pushed up, and it has never been flooded by the sea since then. This uplift was accompanied by strong compressive forces from the southeast, which crumpled the thick wedge of sediments into a series of folds. In places, especially in the southeastern part of the folded belt, the forces were so intense that the rocks were faulted as well, and portions of the crust were shoved over neighboring parts. Toward the northwest the folding was less intense and the layers were merely uplifted into a plateau structure. This, then, was the time of mountain-making when the forces of diastrophism caused the crumpling and faulting of the crust, and when a distorted rock structure was produced which has remained ever since and has determined the development of valleys and ridges for the past 200 million years.

The third and most recent part of the history of the Folded Appalachians has consisted of mountain-carving. This has taken place in a number of separate stages. First, the mountains formed at the end of the Paleozoic Era were worn down to a more or less flat surface, a peneplain (a word derived from Latin meaning almost a plain). Next, the peneplain surface was raised. It is important to realize that there was no further folding of the rocks at this time; all that happened was that the mass of crumpled strata, the tops of whose folds had been removed by erosion in the production of the peneplain, was uplifted so that the forces of gradation again became active. Rivers now are cutting this surface down to a new and lower level, and this process is leaving the more resistant rocks standing up as ridges.

The even ridge tops are a reminder of the former, now uplifted, pene-

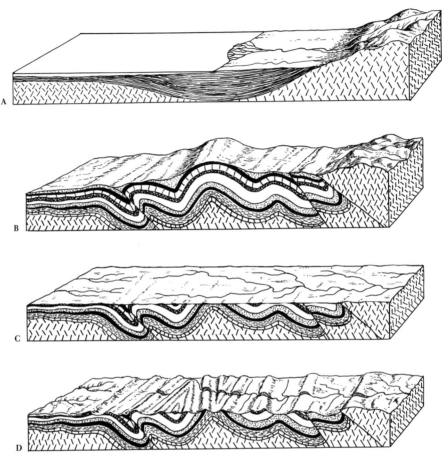

EVOLUTION OF THE FOLDED APPALACHIANS. The block diagrams are generalized. A: Deposition of sediments in the geosyncline. B: Folding and faulting. C: Peneplanation. D: Present day ridges and valleys.

plain. The present mountains are thus really the products of erosion. The area is rugged because the softer material has been removed, the harder being left behind to form the summits. Put in another way, we can think of the higher parts of the land as having come into being only by the hollowing out of the valleys between them. They were not pushed up as we see them today.

If, after the peneplain was formed, the land sank and was covered by a thin coating of sediments and was then again uplifted, streams would have developed and flowed over the level surface of the sediments, unaffected by the underlying uptilted rocks. Such streams would develop courses straight down the slope to the sea, and after cutting through the thin sedimentary cover, would find themselves established on the ridges below and be forced to cut gaps through them. All of this hypothetical cover has now been removed. Its former presence must be postulated, however, to explain the water gaps now existing.

The Appalachian Mountains end in Alabama where their structure disappears under the more recent coastal plain sediments laid down in the deep marine embayment which at one time existed far up the Mississippi River to southern Illinois. To the west of this embayment the same structure and arrangement of the Appalachian rocks appear again in the mountainous area of Missouri, Arkansas, and eastern Oklahoma. Here the Ozarks of southern Missouri and the Boston Mountains of northern Arkansas are formed of eroded horizontal sediments like those of the Appalachian Plateau area. And southward from this area the Ouachita Mountains of south-central Arkansas, which extend into eastern Oklahoma, are composed of folded sedimentary rocks.

In the western part of the United States, from the Rocky Mountains to the Pacific Ocean, the land has been subject to almost every conceivable type of diastrophic movement. This has resulted in many varied examples of mountain types. Each section emphasizes its own brand or combination of the basic movements of folding, faulting, or straight vertical motion of uplift or downwarp.

We have previously discussed the two large plateau areas of this region, the Colorado Plateau at the south and the Columbia Plateau at the north. They, together with an intervening section of fault-block mountains in Nevada, separate the Rockies on their east from two major north-south mountain groups on their west, the Coast Ranges which follow the Pacific border from Canada to Mexico and the Sierra Nevada-Cascade sequence. As has been noted, the Cascades extending from Northern California through Oregon into Washington are composed of a number of volcanic mountains. The Sierra Nevada range, in contrast, is carved out of a large block of the earth's crust which has been uplifted and tilted westward. This range and the smaller Grand Tetons of Wyoming are two of the best examples of this simple type of mountain structure to be found in the United States. Most of the mountain masses in Nevada, western Utah, eastern California, and southern Arizona are fault-block mountains exactly similar in origin to the Grand Tetons and the Sierra Nevada.

Of all these mountain masses, the Tetons probably reveal their fault-block origin best. As we approach the Tetons from the east, we see them growing in stature until we reach Jackson Hole, a flat sage-covered valley twelve miles wide. Standing on the floor of this valley we can look up to the peaks rising a mile above us. They form a magnificent mountain group with a jagged alpine profile. Patches of snow lie on the upper slopes and the gray rock summits thrust upward above the timber line. The lower flanks of the mountains are clothed in deep forests of lodgepole pine and alpine fir, and shimmering aspens are very numerous. Spruce trees grow along the creeks and around the small lakes that lie at the base of the mountains.

The junction between the flat floor of Jackson Hole and the abrupt rise of the mountains is striking. There are no foothills on this side of the range, and if the mountain front is viewed obliquely across Jackson Lake, the line of peaks can be seen more in perspective and its straight north to south trend can be noted.

The range is about forty miles long and about twelve miles wide. Its deeply serrate crest rises into many peaks above 12,000 feet; the highest, Grand Teton itself, has an elevation of 13,000 feet. The western slope of the range is less steep than the eastern and terminates in the Teton Basin, which is somewhat lower than Jackson Hole. On this side the higher peaks are flanked by a series of foothills which grow in elevation as the main bulk of the mountain mass is reached.

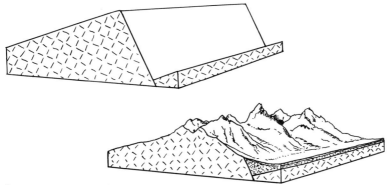

STRUCTURE OF THE GRAND TETONS

The production of the Tetons involved the displacement of a block of the earth's crust with respect to neighboring rocks. The uplift occurred along a fault which parallels the straight eastern margin of the range. The block was uplifted at least 7,000 feet and given a westward tilt. Before being attacked by recent glaciers, the very steep eastern face and the gentle western slope were even more pronounced than they are today.

There are still several small glaciers in the Grand Tetons, none more than a mile long; all of them lie on the east side of the range. These are reminders that much of the present scenery of these mountains is due to glacial action. All the features of a thoroughly glaciated range are exhibited here. There are the deep cirque bites, the jagged arête ridges, and at the top, the horns of rock pointing ever higher. All these combine to present a classic example of glacial destructive action.

The more extensive Sierra Nevada mountains are similar to the Tetons; they have a zone of foothills on the western side, and on the east, a straight fault margin, essentially without foothills.

Much of the Sierra Nevada range is composed of primeval wilderness areas. We have already discussed the Yosemite Valley in connection with glaciation, but although it is perhaps the best known National Park in this region, there are other equally beautiful and striking areas, such as the Sequoia and Kings Canyon Parks. Mount Whitney is located in Sequoia Park; its summit soars to 14,494 feet, and of all the mountains in the United States it is surpassed only by Mount McKinley of Alaska. Deep canyons, rugged terrain, and above all the giant sequoias make this a region of magnificent scenery. It is almost impossible to convey the grandeur of the redwood trees. They tower above tall firs and ponderosa pines, reducing them to insignificance, and the visitor is overwhelmed not only by their immense size but by their incredible age, often measured in thousands of years.

Lake Tahoe, the largest body of water in the Sierra Nevada, owes its origin to subsidiary faulting in these mountains. It lies in a basin, near the crest of this mighty range, produced by faults on both the east and west sides which are roughly parallel to the major boundary fault of the Sierra Nevada block itself. This lake basin is thus a section of the range which was not elevated quite as much as the rest. Lake Tahoe is superbly situated, ringed as it is by mountain peaks. It is nearly twenty-two miles long, its greatest width is twelve miles, and it has a depth of more than 1,600 feet. The lake level is 6,200 feet above sea level and the higher mountains rise to between 9,000 and 10,000 feet in elevation.

EVERY time an earthquake is reported we know that a fault has occurred. The greater prevalence of earthquakes in the western part of the United States in contrast to the eastern is an indication of the more frequent occurrence of faulting, and a reminder that diastrophism is still very active.

In some places faulting involves vertical shifts and in others, primarily horizontal shifting. In the production of fault blocks the shifting is, of course, mainly vertical. In marked contrast, the San Francisco earthquake of 1906 resulted from a horizontal shifting of the crust. It occurred along the famous San Andreas Fault. This is a major break which extends over 600 miles from a point south of the Mojave Desert in southern California to Point Arena, north of San Francisco. The country on the west side of the fault line was moved northward relative to that on the east. Most of the motion along this rift has been horizontal. In 1906 the maximum horizontal displacement, which occurred northwest of San Francisco, was 21 feet, and the maximum vertical displacement was only about three feet. The trace of this fault in the vicinity of San Francisco can be very easily followed on a local map of the area. The location of Tamales Bay and Bolinas Lagoon on the coast to the north of San Francisco, and San Andreas Lake and Lower Crystal Springs Reservoir to the south are all determined by the fault which forms a zone of weakness, more readily eroded than material on either side.

Careful surveys have shown that there is currently a yearly shifting of the crust of about two inches in the San Andreas fault area. The country to the west of the fault is still moving northward relative to that on the east side. There is not a continuous shifting along the fault; this relative motion must result in a twisting of this part of the land. In time, as twisting increases, and when stresses have been built up sufficiently enough to overcome the frictional drag along the fault, movement will occur with a jerk and all accumulated stresses will be relieved in a single motion. This slow buildup of stress in the crust and the sudden readjustment along a fault, resulting in an earthquake, is probably the method by which most such crustal readjustments occur. In the production of such a major feature as the Sierra Nevada mountains, movement along the boundary fault undoubtedly occurred many times before the total adjustment, measured in thousands of feet, was attained.

In driving through western Utah and Nevada to the Sierras, we may have noted that many of the fault-block mountains of this area have been well dissected. Now only the remnants of these mountain masses remain, protruding through the tremendous amount of material which has been washed from their summits into the intervening basins. A typical basin now presents a broad, slightly rounded bottom, floored at its center with hundreds or thousands of feet of recent sediments. This is the Basin and Range area, a great region that extends from the edge of the Columbia Plateau in the north to the Mexican border on the south. Its eastern boundary is marked by the Colorado Plateaus, and at its western limits lie the Sierra Nevada.

This area is a region of little rainfall with a semiarid climate. Vegetation cover is spotty, and large areas are essentially devoid of growth. There are very few perennial streams, and gullies and canyons have water in them only immediately after a rainfall. It is, furthermore, a region of interior drainage, that is, most of the precipitation never reaches the sea, but either sinks into the ground or evaporates.

When the traveler in this region consults his map he can note that some of the rivers vanish without reaching a body of water, and that others may flow down to the low parts of the basins to form a lake, which may disappear in the dry season, and whose bed may be given some such label as "dry lake," "soda lake," "alkali flat," or "borax lake." These ephemeral playa lakes commonly leave "salt" incrustations when they dry out, and the type of salt may vary from basin to basin. Since water never leaves such a lake except by evaporation or by sinking into the ground, there is a concentration of salt continually brought in by the intermittent streams. A few of the larger of the salt lakes persist throughout the year. The Great Salt Lake in Utah is perennial, but with a variable level. After a series of dry years the water level drops, and after heavy rainfall it rises. In 1937, for instance, the visitor had to travel hundreds of yards from the bathhouse at Saltair to reach the water, while in 1953 the water was a couple of feet deep at the bathhouse.

During the Ice Age there was a great deal more precipitation than at present in the Basin and Range area, and the Great Salt Lake was many times larger. At that time it covered the Great Salt Lake Desert and spilled over to include Utah Lake and Sevier Lake, which are now isolated lake basins. This Ice Age lake, Lake Bonneville, left as evidence of its former extent many wave-cut cliffs, terraces, deltas, and sand bars. As we stand on the shore of the present Salt Lake we see around us long lines of terraces on the neighboring hills which show its former extent. The cliffs behind the terraces are wave-cut features and mark the edge of the lake in former times. Some such cliff remnants are especially conspicuous on the flanks of the Oquirrh Mountains just to the south of the Great Salt Lake, and the Bonneville race track is laid on the beautiful packed sand floor of the former more extensive lake.

Driving across the Basin and Range Province from Provo, Utah, to Ely, Nevada, on U.S. Routes 50 and 60, we are in a typical region of fault block mountains and basins. In Utah, the emphasis is on the very wide and extensive flats in the floors of the basins. In places the flat floor extends right up to the eroded mountain face itself. These level floors are the beds of former lakes, some of which were parts of the extinct Lake Bonneville.

Such flat floors which directly abut the more precipitous valley side have an origin exactly similar to that of the flat part of Yosemite Valley, the Lake Agassiz flat area, and the level valleys of the Finger Lake Region of New York State; they represent the level top of layers of sediments which have accumulated in lakes which are now extinct.

On its way further west from Ely to Mono Lake in California, Route 6 crosses a number of basins and their separating ranges. In this area the basins on the average have a very gentle slope downward from the encircling mountain fronts to a low spot at the center, in which, as we have mentioned, there may at times be an ephemeral lake. This is a dry and desolate country with far reaches to the mountain-ringed horizon. The route consists of straight stretches of road, often over twenty miles without the slightest turn, across sage-covered flats interspersed with short climbs and a winding course over the almost buried mountain ranges. Most of the valleys have a central narrow flat floor, which marks the dried-up lake bottom, from which gentle slopes rise to change abruptly into the steep rocky flanks of the mountains.

On a hot dry summer day "dust devils" whirl over the valleys and queer mirage effects play over the simmering land. They help to break the monotony of the drive over the interminable straight stretches of road. As one starts dropping down the gradual decline at one side of a valley, one can look across to the opposite side in the distance, and realize that the gentle slopes are composed of a whole series of coalescing alluvial fan deposits which were spread out by the infrequent mountain streams, when they emerged from their steeper mountain channels.

Contrary to widespread belief, the wind, even in this arid region, moves a very small amount of material, compared with running water.

In addition to the few perennial and the many intermittent streams, the principal agents of erosion in arid basin areas are splashing rain, rill wash, and sheet floods, where the water runs down slight slopes as a sheet, rather than concentrating in channels. Since there is very little vegetation to hold the water back, a tremendous amount of erosion in short periods of time is accomplished during the rare times of rain and flood. This is the region of flash floods where a camper who has unwarily pitched his tent in a dry stream bed can find himself suddenly displaced by a raging torrent of muddy water. But such a flood does not last long. The water is soon either evaporated or absorbed by the dry ground.

LEAVING the arid stretches of the Basin and Range province, with its ragged worn-down peaks and endless alluvial flats, we move on now to the best-known mountain range in the country, the Rockies. What a contrast there is between Death Valley, the most desolate of all regions in the Basin and Range province, and the Bighorn Mountains of Wyoming! The Big-horns are an isolated part of the Rocky Mountain chain located about half-way across the state near the Montana border. If we approach them across the flat Powder River basin on the east, heading for Sheridan, we see them looming ahead of us, blue-green and timber clad, and we realize that our

SHEEP MOUNTAIN, WYOMING.
An anticlinal mountain fringed with hogbacks.
Barnum Brown

road will probably have to climb them in a series of switchbacks. Our progress is not likely to be very rapid, and we shall have plenty of time to look at the geological structure of these mountains.

The Bighorns are an excellent example of a simple folded structure. They have been carved out of essentially one major wrinkle of the crust of the earth, over one hundred miles long and twenty to thirty miles wide. Shortly after leaving Sheridan our road starts to climb the very abrupt east face of the mountains and here we can see how a series of sedimentary rocks is tilted toward the east. The more resistant layers stand out in bold relief as hogbacks, while the weaker layers have been removed. The harder layers thus now form a series of north-south ridges which parallel the mountain front. These lower ridges are composed of very colorful red and white sediments, in marked contrast with the green grass. Higher up where the darker evergreens begin to appear, there is a very noticeable white resistant layer, whose edges are visible at the sides of the gullies which have cut into the mountainside. Between the gullies the resistant white sediments stand out as remnants, apparently plastered onto the slopes.

The road up is very well graded, and in a short time we get some spectacular views over the plains to the east. As we continue to climb, the sediments can be seen to dip less steeply, and at the top they are essentially horizontal. Here, however, they are preserved only in patches, especially in the higher parts, as most of the rolling land is underlain by granite bedrock. This is a thickly forested area, one of the heaviest timber stands in Wyoming. Our road runs for miles through woods of sweet-smelling pines, spruce, and fir, with here and there open parklike areas carpeted with grass and wild flowers.

On going up from the plains we have cut backward in time, and on the summit of the mountain we stand on rock which once formed the basement on which the sedimentary layers were deposited. This granite basement with its overlying sediments was pushed up to form a great arched fold, or anticline. Now erosion has removed the top layers of the arch to reveal the granite. Farther toward the west the sediments dip down and disappear under the Bighorn Basin which adjoins the mountains on the west. At the northern and southern ends of the Bighorn Range, erosion has not cut deeply enough to penetrate through the sedimentary cover, and thus the granite is not exposed as it is in the middle parts.

As the road leaves the Bighorns going west it winds down Shell Creek Canyon. Here Shell Creek has cut deeply into the westward dipping sedimentary rocks to expose the coarse underlying granite. If we park our car long enough to look at the rocks exposed in one of these cuts, we can read an interesting chapter in the history of the mountains. The lowest rock in the section is granite, marked with intersecting horizontal and vertical joints. Curiously, directly on top of the granite is a layer of sandstone which is composed of minerals which were weathered and broken off from the granite. Granite, it will be remembered, is always formed deep in the earth, so there must have formerly existed a great thickness of some rock, no longer present, into which the molten granite intruded from below and cooled. Then, some time in the following ages, the restless forces within the earth raised this part of the crust, and through long succeeding ages it was weathered and eroded until the rock overlying the granite completely vanished and much of the granite itself was broken down into the minerals which composed it. Next, conditions changed from erosion to deposition, probably through subsidence of this part of the earth. The minerals that had been eroded from the granite were laid down as sediments and, through the following ages, became hardened into the sandstone we now see. Since that chapter in earth history, the crust has again been uplifted essentially into a single fold, and once more erosion has uncovered the granite and is continuing its interrupted job of cutting still more deeply into it.

Such an erosion surface as that between the granite and the overlying sediments is known as an "unconformity," and always implies a time of erosion followed by the deposition of sediments. We have already discussed the implications of the unconformities seen on the lower walls of the Grand Canyon.

From a lookout point on the Shell Canyon road, about eight and a half miles before reaching the town of Shell, the wide flat Bighorn Basin can be seen to stretch for miles away to the west and in the far distance, over 60 miles away, some snow-covered mountains, which lie between Cody and Yellowstone Park, are visible on a clear day. In the immediate foreground hundreds of feet of alternating layers of buff- and red-colored sediments can be noted overlying the granite, which is uncovered nearer the bottom of the Canyon. The sediments and sediment-granite contact dip gently westward. The slopes here are dotted with sagebrush and small conifers and junipers.

At the exit from Shell Canyon there is a very colorful red sandstone, and farther out, in the Bighorn Basin proper, there is a touch of the Painted Desert scenery in the barren, gently rolling clay hills with their red, purple, and gray colors, and the shallow creasing of their slopes by small gullies.

From a few miles away one can get a fine view of the Bighorns we have just crossed, and can notice how the dip of the sediments abruptly increases as the mountain front is reached, and how the line of hogbacks is formed, composed on this side of resistant layers dipping westward.

The scenic details in the southern Rockies in Colorado are very similar to those of the Bighorns. Hogback ridges fringe the older denuded cores of granite and metamorphic rocks. On the eastern border of the Front Range there is an especially fine collection of hogbacks formed by resistant sedimentary layers which dip toward the east. The dip may vary, but the hogbacks extend almost continuously for the full length of this part of the Rockies.

At Boulder, Colorado, we can get a very good idea of the typical arrangement of the rocks. The westward trip from the plains east of Boulder into the mountains starts on the rolling, slightly eroded plains of eastern Colorado. The rock layers as shown in road or stream cuts, or from drilling information, are essentially horizontal and many thousands of feet

thick. The plains come to an abrupt end as Boulder is approached. The deeper sedimentary layers are turned up and the ragged edges of the resistant rocks show up as hogback ridges. They emerge through the veneer of more recent sediments washed down from the Rockies.

Near Boulder the westernmost hogback ridge is composed of a coarse pebbly sandstone which lies directly on the granite. The granite here happens to be slightly less durable than the sandstone which thus stands up as a ridge superimposed on the principal granite mass. The sandstone has a scalloped appearance due to the cutting of stream valleys across its contact with the granite, and it looks as though a series of giant flatirons were leaning up against the higher mountains.

West of the hogback zone the main mass of the granite and metamorphics of the Rockies begins. The surface between granite and sandstone, now uplifted in the Rockies to a height of over 14,000 feet, underlies the plains east of us at depths of thousands of feet. Here, accordingly, one must think of the originally horizontal granite and sedimentary rock contact, which is exposed at Boulder between the "flatirons" and granite, as continuing eastward underneath the plains, and as once having continued westward up and over the present high granite peaks of the Front Range. It represents the eastern side of one tremendous upfold, or anticline, the western side of which comes down again in North Park miles to the west.

At the crest of the Front Range in the Rocky Mountain National Park on the Trail Ridge Road, a high point of over 12,000 feet is reached. Here the land is surprisingly flat. We are on a rolling upland surface, which is in very marked contrast to the rugged steep-walled canyons cut into it from all sides. This upland area of low relief is a remnant of an erosion surface produced when the Rockies were much lower in elevation than they are now. Since that time an uplift which slightly arched this old erosion surface has allowed streams, and later glaciers, to cut into the uplifted land mass, producing the canyons, cirques, and steep cliffs which we see everywhere. The arching of the peneplain marks a late stage in the diastrophic history of the region. The initial folding which resulted in the steep hogback dips was far earlier. The history here is very similar to that of the Appalachians: folding and erosion followed by renewed uplift and erosion.

This Rocky Mountain country has a beauty which is a great attraction for the tourist. As we drive through it, however, we realize that there must have been something else to draw the early pioneers to such a rugged land. In the Front Range in Colorado there are many largely deserted mining towns, derelict buildings, mine dumps, and mine openings. From some lookout points it is possible to see a dozen or more prospect holes, with a pile of debris at the exit. They look like giant woodchuck holes from a dis-

THE BEARTOOTH MOUNTAINS, MONTANA. Note the many cirques and the subdued rolling topography of the uplands. This upland surface marks the level of a former peneplain, which has been elevated and partially destroyed by river and glacial erosion.

Courtesy Montana Highway Commission

tance, scattered as they are all over the hillsides. Around the Boulder area there formerly was extensive mining of gold, silver, copper, and tungsten, and more recently of uranium.

The early prospectors knew that where there are mountains there is likely to be valuable ore. Ore deposits are commonly associated with mountains and the intrusion of igneous rocks. Such deposits usually take the form of veins produced by hot emanations of gases and liquids derived from the cooling magma.

South of the Front Range of Colorado the Rocky Mountains extend as far as the end of the Sangre de Cristo Range in New Mexico. If we find ourselves at the southern limit of the Rocky Mountains, we might bear in mind that we are at the end of a great range which extends from this point in New Mexico all the way up through the United States and well into Canada. Its most spectacular and rugged scenery, in fact, lies in its northern reaches. The Canadian National Parks of Jasper and Banff are well known for their glaciers, their snow fields, beautiful lakes, and rocky peaks, and we find the same alpine scenery in Waterton Lakes Park which lies side by side with our own Glacier Park. If we have visited the Tetons before we go to Glacier, we are prepared for the type of glacial scenery we shall find there, but at Glacier Park everything is on a much grander scale.

Here in the northern part of the Rockies, many of the rock layers in addition to having been closely folded and uplifted were extensively faulted. Parts of the crust have been literally pushed over other parts for distances of many miles. Chief Mountain, Montana, situated on the boundary of Glacier National Park, is an eroded limestone remnant of a slice of the earth which was thrust up and over adjacent parts for a distance of more than ten miles.

WHILE all present-day mountains, outside of the most recent volcanoes, are, in their scenic details, really erosional, we must realize that generally our view of scenery is circumscribed. It is often very difficult to appreciate the large scale of the original shape of a landform produced by diastrophism. The folded and faulted parts of the crust can be understood properly only with the help of our imagination, aided perhaps with maps and geologic sections.

The basin-and-range type of structure results from vertical forces. In the case of very simple dome-like folds such as the Black Hills in South Dakota, the diastrophic force must also have been vertical and acted like a finger under the crust which pushed the layers up into a little bulge. In the Black Hills the high parts of the dome have been worn away to expose a granite core, which is now surrounded by a ring of hogbacks dipping from the center of the Hills.

The production of folded structures, on the other hand, necessitates side pressures, or forces working parallel to the earth's surface. The Bighorn Mountain-type of folded structure is found in many regions, with local variants. In other places the structure may consist of a series of folds one after the other, that is, the crustal layers may be crinkled into alternating up-and-down folds, or anticlines and synclines, such as we find in the Appalachians. And if the crumpling forces have been intense enough the rocks may be faulted sometimes quite considerably, as we have seen at Glacier Park.

Scenery in the mountains is due mostly to running water, but we owe many of the more rugged features of mountain landscapes to the enormous power of moving ice, such as the jagged sawtooth type of mountains, like the Grand Tetons, the Alps, and the Himalayas. Streams produce a smoother and rounder type of mountain silhouette. For instance, the smoothly carved slopes of the White Mountains, exclusive of the more superficial, glacially produced cirque bites and oversteepened sides of the U-shaped valleys and the rounded forms of the Catskills and the Great Smokies, are typical stream-produced features; when streams work on folded rocks with alternating bands of hard and weaker layers, as in the

Appalachians, they produce a series of ridges with intervening valleys.

In discussing the making of mountains, the term *crust* of the earth has been used, without anything being said about its thickness or general properties. The crust is the surface layer of rocks, with an average thickness under the continents of from 20 to 25 miles. There is a notable difference between it and the material directly underlying it with respect to the speed with which earthquake waves are transmitted through them. Furthermore, in the crust itself there are lateral variations in the speed with which earthquake waves travel, owing to differences in the type and density of the rocks of which it is composed.

It is known that the rock masses that make up continental lands are composed of material of lighter weight than that underlying the ocean basins. The continental blocks are made of rocks with a density roughly of granite, the oceanic rocks are more like basalt in density. The relationship of continent to ocean basin can, in a sense, be likened to that of iceberg and sea. The floating iceberg is composed of lighter material than that of the water, and the land areas can be considered as "floating" on a denser lower rock. The term float must, however, be qualified. The rocks under the continents are not liquid in the usual sense of the term, but will yield slowly under pressure while retaining their solidity somewhat in the way that glacial ice does in flowing over the land.

What causes parts of the crustal layers to wrinkle into folds and other parts to slip along faults? Various explanations have been offered, among others that there are convection "currents" in the denser material below the crust, perhaps set in motion by greater radioactive heating at one point than at another. Such currents, it is postulated, may cause a frictional drag on the overlying, more brittle crustal layers, which in turn may yield by either folding or fracturing. It has been suggested that diastrophism is the result of the earth cooling and thus shrinking, and that mountain chains are analagous to the wrinkles on a desiccating apple. However, doubt has been thrown on this suggestion since the temperature of the earth has been shown to have been fairly constant through long periods of geologic time.

Whatever the cause, mountain belts do exist. The high places of our land have a special, compelling attraction. Instinctively, we quicken our pace when we approach a mountain range. The peaks seem to beckon to us and we can scarcely wait to reach the foothills and begin to explore the heights. Sometimes we are drawn by the lure of breathing crisp mountain air, or by the exhilaration of being on top of the world, but above all we know that in the mountains we shall find magnificent scenery.

MEANDERS OF THE WHITE RIVER, INDIANA   *Photography-CSS-USDA*

CHAPTER 8

THE major force in the production of scenery is water running on the surface of the land. Of all the agents of erosion, including ocean waves and currents, the wind, ice, and underground water, it stands preeminent; it moves more material a greater distance than all the others combined. Water is responsible for the most obvious scenic elements on the land—the hills and the valleys. As a general rule, hills are residual features left after the

Water on the Land

valleys between them have been carved out by stream action. There are exceptions, of course, such as glacially formed drumlins, eskers, and kames, the wind-formed dunes, and the hills due to the build-up of volcanic debris, but these occupy relatively small parts of continental areas.

Every year about 100,000 cubic miles of water is evaporated from the earth's surface. A little over one third of this water falls on the land; of this third, much more than one half is evaporated, the annual runoff which eventually reaches the sea being only about 9,000 cubic miles. Most of the remaining water has some erosive effect during its short life on the surface, before it disappears back into the air or temporarily underground. Even in desert areas, such as those found in Nevada and Utah, where no precipitation ever returns to the sea, the work of water is of vital importance. As we have seen, it is so effective that the dominant scenic forms in many deserts are those ascribable directly to water, whether flowing in channels or spread out over the land as sheet floods.

Some water makes at least part of its journey back to the sea in a frozen state. In the polar regions of today it may make its complete journey in this fashion, whereas in the middle latitudes it may make the trip down mountains as part of a glacier, and at lower elevations as a liquid.

Before it is channeled rain runs off a slope as a sheet of water. At this stage it can move a great deal of fine material under proper conditions. The surface of a recently ploughed and smoothed field during a heavy rainstorm will be seen to be alive with sand grains jumping and shifting under the impact of the raindrops. And if the field is on a slope a sheet of muddy water carrying much sand can be seen moving downhill. Very quickly the sheet of water concentrates its flow into slight depressions which become filled with more rapidly moving water, which in turn picks up and moves more material, thus cutting a still deeper channel into the loose soil. The rapidity with which topsoil can be washed away is startling. Much fertile soil has been lost to the sea as a result of poor farming methods. The way to avoid such soil erosion is obviously not to allow the runoff to become concentrated, but to slow up the rate of flow so that as much water as possible is allowed to seep into the ground. In contour plowing the furrows are made along the slope, not down it, to check sheet flow and to help prevent the carving of gullies. By far the best way to prevent soil erosion is by planting an earth-hugging vegetation cover, which will slow up the rate of runoff and help to hold the soil in place by its network of roots. Furthermore, measures to check the removal of soil obviously help also to prevent floods.

The geologic work of water as it returns to the sea has two aspects. It washes away loose material and uses such material while in transit to scratch and break up the bedrock over which it flows.

This dual nature of the work of a stream, the abrasive and the transporting, is necessary for the cutting of the typical youthful valley. The gorge of the Yellowstone River just below Lower Falls illustrates this excellently. Here the river has cut into a thick sequence of lava flows and uncovered a veritable rainbow of colored rocks, of yellows, pinks, and browns. The two walls of the valley slope uniformly down until they meet at the stream. The cross profile of the valley is in the shape of a V, with the river filling in the very bottom. It seems obvious that the material from this V-shaped cut must have been removed by the river. If, in imagination, we go back to a time before the valley was cut, we can picture the ancestral Yellowstone River when it had just produced a shallow groove into a relatively flat area, long before the falls came into existence. Just as soon as this groove was cut it left two slopes exposed to view, rocky slopes which could be attacked by the forces of weathering and decay and down which the crumbled rock could slide or be washed into the river. The stream then used this material to scour out its channel still deeper and create still more valley wall to be exposed to weathering and erosion. Two processes thus united to form the valley. The upper walls of the valley have retreated the farthest because they have been exposed to the forces of rock decay for the longest time.

LOWER FALLS OF THE YELLOWSTONE RIVER, WYOMING. The falls and rapids and V-shaped valley profile are characteristic of youthful streams. The falls are located where the river leaves hard volcanic material for softer volcanic material downstream.

National Park Service Photo

The Yellowstone Gorge has resulted in a symmetrical profile because all the rock layers here have more or less the same resistance to weathering. In the case of the Grand Canyon of the Colorado the rock layers differ in resistance and thus the rate of retreat of the various levels of the valley wall is different and a steplike profile is produced. Wherever the breaking down of the sides of a valley is slow compared with the rate of river abrasion the walls will be steep and may be almost vertical. Such a valley may vary in depth from a few feet up to several hundred feet. Ausable Chasm in the Adirondack region of New York State is a good example of this type, where the top of the valley is not very much wider than the floor.

The V-shaped transverse profile, whatever its modification—steplike, steep, or gentle—results as long as a stream remains active enough to remove all the material which has slid down the valley walls or been brought by its tributaries. Under these conditions the river will flow on bedrock and continue to cut its channel ever deeper. Such a stream is said to be young. The Yellowstone River, the Colorado River, at the bottom of its mile-deep canyon, and a small pasture brook are all young; they are striving to cut deeper. By far the greatest number of small and medium-sized streams, as well as a number of the largest ones in the world, follow a youthful winding course down valleys with such a V-shaped cross profile.

If part of the rock of a stream bed is softer than another part it will be removed more easily. Whenever the softer rock is downstream, falls or rapids may result. In the case of Yellowstone Falls, for example, the river drops from harder volcanic material onto softer volcanic rock downstream.

Waterfalls and rapids are transitory features of a river. It is at just these places that the stream has most energy of motion, and that the tools in the form of sand and gravel which it carries scrape with the greatest force. At the base of a waterfall such as Niagara, which has a cap rock of resistant limestone, the turmoil of the water stirs up any material in the stream bed and by the grinding eddying action actually erodes upstream under the falls as well as cutting a "plunge pool" in the river bed itself. Thus, periodically the support of the cap rock is removed and it collapses. Such has been the history of the Niagara Falls. Every few years the resistant cap rock has collapsed through the undercutting of the softer rock below it. The present Niagara Gorge is the result of a retreat of the falls for a distance of 6½ miles in the last few thousand years.

The plunge pool at the base of a major waterfall may be a very large feature. For example, in the Grand Coulee, an abandoned channel of the Columbia River, there is a plunge pool eighty feet deep and a half a mile across, formed when the river fell over a four-hundred-foot cliff. Similar, but smaller, holes cut in the bedrock of a youthful stream are very common, and are produced by swirling water, which, using pebbles and boulders as tools, literally drills such holes in the bedrock. Such "potholes," as they are called, occur in various shapes and sizes. There is an especially fine pothole called The Basin, a few miles south of the Old Man of the Mountains in Franconia Notch in New Hampshire. This is a hole about twenty feet in diameter which the Pemigewasset River hollowed out in the rock of its bed with the aid of swirling sand and gravel. The Basin was probably largely cut by meltwaters from the retreating glacier farther north, when the amount of water in the Pemigewasset was much larger than at present. Here at the Basin beside the large pothole, the bedrock has been smoothed and scoured into streamlined shapes. Because of slight differences in hardness, a delicate relief pattern has been etched and carved into the bedrock. Where the stream makes a bend, the rocky banks have been rounded and slightly undercut by the torrent. In the production of potholes a stream must always use its load of sand and gravel as abrasive

POTHOLES IN THE BED OF THE PEABODY RIVER, PINKHAM NOTCH, NEW HAMPSHIRE.

Photo by John Shimer

tools; without these tools a stream is practically helpless as an agent of erosion.

Both Niagara and Yellowstone Falls were developed in the course of their orderly growth as youthful streams, cutting down into rocks of diverse hardness. The falls and rapids of the Potomac River at Washington, D.C., were similarly produced. Here the river leaves the harder rocks of the Piedmont area of the Appalachian Mountains to flow over the softer rocks of the Coastal Plain. The Potomac is but one of many rivers which crosses this boundary line and which at this point in their paths have very noticeable rapids. This boundary is thus appropriately called the Fall Line.

Falls and rapids also occur in a number of other situations. We have seen that falls result when a valley glacier deepens a main valley to such an extent that tributaries are left hanging. Steep slopes are formed also by the rapid downcutting of a youthful stream so that tributaries coming in from the side must fall down the valley wall. The small side streams descending very rapidly down the walls of Grand Canyon to the Colorado River are excellent examples of this type. If the Colorado were not so much more powerful in cutting its channel than its intermittent tributaries, the Grand Canyon would of course look quite different today. This is a somewhat unusual situation because generally tributaries keep up fairly well with the rate of downcutting of their principal stream and enter at grade.

Young streams are apt to change level and rate of flow often, and very rarely indeed are navigable. This is in marked contrast to the "mature" and "old" streams which have gentle gradients and provide some of the major waterways of the world. A stream reaches maturity when, by cutting the level of its channel, it has so reduced its gradient and thus the velocity of its water that it has only just enough energy to transport its load of mud, sand, and gravel, not enough to scour out its channel any deeper. At this stage it starts to cut sideways, making a flat surface on which it flows. The V-profile is abandoned and a valley floor is slowly developed on which the stream may meander. Generally such a stream is confined to a channel on this flat surface, but at times of flood the whole valley floor, or flood plain, may be inundated. The sides of such a mature valley continually recede farther and farther apart, the valley walls being slowly weathered and washed away and periodically eaten into here and there by the meandering stream. The meanders of such a stream never stay in place for very long but move from side to side and migrate down the valley. This motion of the meanders may not be obvious in the span of a few years but during the course of a few hundred years there are apt to be important changes in the path taken by such a stream. In the early days of this country long stretches of both the Mississippi and Missouri Rivers were used as state boundaries, and modern topographic maps will show that in many places now the rivers have moved miles from their former positions as marked by these early boundaries.

The change from youth to maturity in the growth of a stream has far-reaching scenic effects. The ungraded, rocky river with falls and rapids and steep walls eventually changes into a graded, slower-moving, meandering stream, carrying sand and mud primarily, and flowing over a flood plain composed of its own deposits. The valley floor becomes in time much wider than the meander belt itself, and is bordered by a zone of bluffs. At this stage the stream may be considered to have reached "old age." Most of the mature and old streams of the present time are found to be flowing on rather thick flood plain deposits; in places along the Mississippi and Missouri Rivers the depth to bedrock is several hundred feet.

As mentioned previously, the meander loops are not stationary features but move over the flood plain. They shift generally outward and down-

THE WHITE RIVER NEAR EDWARDSPORT, INDIANA. Note the many abandoned channels of this meandering river, which are now marked by ox-bow lakes. The pattern of many of the fields is broken by crescent-shaped white patches, marking the deposits of sand on the inside of former meander loops. *Photography–CSS–USDA*

stream. A stream flows most rapidly on the outside of such a loop, and it is here that the water hits the bank and cuts it away. On the inside of the meander loop the water flows less rapidly and sand bars are deposited. Thus in time a meander loop will transfer itself across its own flood plain deposits by eroding on the outside and building up on the inside. The deposits of a flood plain consist of a whole series of such sand bars plus a periodic thin coating of mud laid down when the river overflows its banks. Meander migration takes place with relative ease because the material which the river transports and deposits is so readily moved. Each meander follows the same procedure in shifting its location, with the result that if viewed throughout some hundreds of years the stream would seem to be writhing down its channel. Periodically a meander loop may become so elongated that the stream at a time of flood may jump across and avoid the long journey around the loop and thus leave it as a crescent-shaped lake on the flood plain. Such "ox-bow" lakes are a common feature of many meandering streams which have been in a mature stage for a considerable length of time.

The flood plain of the Mississippi River below Cairo, Illinois, varies in width between 20 and 60 miles, and is one of the classic localities of the world for the illustration of a great variety of flood plain features. The meander loops are large, owing to the size of the river, and most features are best seen from the air. From such a vantage point, ox-bow lakes, abandoned channels, and abandoned crescent-shaped sand bars laid down on the inside of the meander loops can all be clearly noted.

Rivers transport their load in three ways. Soluble material such as calcium carbonate and various salts are carried in solution. The finer mineral and rock fragments are carried in suspension, being kept above the river bed by the turbulence of the flowing water. Coarser material is pushed or rolled along the bed of the stream, and is generally moved only at times of high water.

Deposition occurs whenever the rate of flow of a river is decreased. As long as a mountain stream is confined to a channel with a steep gradient it can carry a large load of coarse material, but when the gradient lessens or the water flows out over a larger area the load is dumped often into a fan-shaped deposit. Such alluvial fans associated with intermittent mountain streams are found very well developed in the Basin and Range country of the West. If the rate of flow of a stream is checked on entering a standing body of water a delta results. Temporary deposits such as sand bars are constantly being made every time the flow of a river diminishes, to be moved farther at the next time of flood waters. The deposits on a flood plain are increased every time a mature stream overflows its banks and spreads out with its load of mud and sand. At such times when a stream starts to overtop its banks the rapid flow in the main channel is abruptly slowed and much of the coarser material in transit may be dumped to produce a low ridge near the river, a natural levee. Sometimes when a river has inundated its flood plain the natural levees can be seen as two meandering parallel ridges protruding slightly above an otherwise flooded land.

Running water is an excellent mechanism for sorting coarse materials from fine, sand and gravel from mud. Effective sorting is also performed on sands of different density. Dense, heavy sand will drop before sand composed of lighter-weight minerals. In this fashion some of the heavy useful minerals have been sorted out and deposited in the lower layers of stream channels. Gold, diamonds, tin, and platinum, among others, have all been concentrated in this way by streams into what are called placer deposits. The first gold found in California, which led to the gold rush of 1849, was placer gold which the rivers flowing westward from the Sierra Nevada

VERTICAL JOINTS IN SANDSTONE, AUSABLE CHASM, NEW YORK. The horizontal lines mark bedding planes. *Photo by Kirtley Mather*

Mountains had concentrated along their channels. The gold was originally disseminated in quartz veins in the rocks of the mountains and was released to be washed down into the stream channels when the bedrock in which the veins occurred was broken into small fragments by weathering.

As one looks at "solid" rocks it is sometimes difficult to realize how they are constantly changing and how important weathering is in the making of scenery. No rock, however resistant it may seem, can long withstand exposure to conditions at the earth's surface. Rock used for building, such as limestone and sandstone, will, in a few generations, start to peel and flake off or become pitted by the removal of some of the less well-cemented parts. It is well known to geologists that to get fresh hard rock one must dig below the discolored rotted material near the surface. The alteration and breakup of rocks is greatly accelerated by the universal presence of joints and, in the sedimentary rocks, of stratification planes also. These cracks and planes afford access to the center of rock masses for the entrance of oxygen, water, and carbon dioxide, the most important chemical agents in weathering.

Rocks may disintegrate mechanically or decompose chemically. Water along joint or stratification planes expands on freezing and causes disruption. Such shattering is of course most effective in the high latitudes or elevations where melting followed by freezing occurs relatively often. When plant roots probe down into joint cracks and expand they also provide a disruptive force. Furthermore, heating and cooling a rock surface hastens its disintegration, as do the mechanical forces accompanying change in volume during rock decomposition, such as those discussed in the exfoliation at Half Dome in Yosemite Valley.

A rock decomposes when one or more of its mineral constituents dissolves or changes into another compound. In either case the initial hard resistant fabric of the rock is destroyed and a pile of loose mineral and rock fragments results. Sometimes the chemical changes are quite complex, but the result is the same, the reduction of a rock ledge into fragments. Weathering is geologically a relatively rapid process. In a few hundred years the average rock will show definite signs of wear.

All soils are fundamentally composed of broken and chemically altered rock fragments. In some places where the undisturbed soil cover has been suddenly exposed, for example, on the side of a fresh road or river cut, one can trace how a humus-rich layer of topsoil changes with increasing depth into a rotted, weathered parent rock, and deeper still into hard fresh un-weathered rock. In contrast, transported soils, such as glacial soils, the material of flood plains and wind-blown material, may have mineral and rock fragments totally unlike the bedrock on which they rest.

Weathering and stream erosion are constantly at work striving, it seems, to cut the land down to sea level. If for a time diastrophism and igneous activity were to cease in a region, the forces of gradation would produce a land sloping very gently upward from the seacoasts, drained by sluggish, meandering streams—in other words, a peneplain. No mountain would appear on the horizon; the land would rise gradually to low, rounded elevations, situated well away from the major river courses. Erosion would be taking place, but at an incredibly slow rate, and as the elevations lowered, the rate would be still less. Perhaps in places a more resistant part of the land might persist for a time as a hill. Such a remnant protruding above a peneplain surface is called a "monadnock" after a mountain of this type in southern New Hampshire.

Mount Monadnock rises above a peneplain which formerly existed in New England but which now has been largely destroyed following a general uplift of the land and a rejuvenation of all the streams. The former peneplain can be noted only in the generally accordant height of most of the lower hills. If we imagine all the land between the small hills filled in, we would see a generally flat surface, above which Mount Monadnock would rise.

The surface of the United States is being lowered at an average rate of one foot in 9,000 years. The Colorado River every year transports through the Grand Canyon 25 million tons of debris, and the Mississippi River dumps between 600 and 700 million tons of material annually into the Gulf of Mexico. These amounts are far too large to have any real meaning to us. To appreciate the quantities involved, however, we might imagine the material carried down by the Mississippi River in the past hundred years as being spread out over Manhattan Island; it would cover the island to a depth of nearly 1,000 feet, leaving only the tops of the Empire State and the Chrysler Buildings to rise above the mass of debris. This mass of material comes from all over the drainage basin of the Mississippi and is channeled into the main river by its tributaries and subtributaries, which reach out into the country on all sides.

Many scenic features are of such a large scale that one needs a bird's-eye view to appreciate them, as for example the patterns that rivers make on the land. To appreciate them properly, the traveler must abandon his car and look at the land from the vantage point of a plane or a mountain-top, or he may even do a bit of armchair traveling over a map. By any method the proper perspective can be attained and the differences between patterns can be noted. An individual river has a very irregular path. It will twist or meander, and it will never be straight for any great distance. When we look at groups of rivers, however, we see that the patterns which they make may be very different in various parts of the country.

On a map of the United States by far the most obvious river system is that of the Mississippi with its major tributaries of the Ohio and the Missouri. This system as a whole has a very clear dendritic appearance, the same pattern of the valleys which were drowned at Chesapeake Bay. This is the characteristic stream pattern which forms in areas of horizontal sediments, such as plains and plateaus, as well as in regions of homogeneous igneous rocks.

To understand the reasons which underlie various river patterns one must bear in mind two things: For geologic ages diastrophism has been raising and twisting the crust of the earth, and at the same time rivers have been wearing down the land areas and, with very few exceptions, cutting valleys where the rock is most easily weathered and eroded. The process is selective; of the stream channels which are initiated all over a land surface, only those which are located on the least resistant material can cut their channels rapidly and thus outstrip all others. If there is essentially no difference between one path and another the dendritic pattern results, and the placing of the various streams appears to be a matter of chance. However, wherever there is any systematic differences in rock resistance the stream pattern will show it.

The roughly parallel ridges and intervening valleys in the coastal area of Maine were developed by the erosion of a land of metamorphic rocks in which there were alternating weaker and more resistant types with a north-south trend.

Where sedimentary rock layers have been folded, bands of rock of diverse hardness which are brought to the surface control the location of stream valleys with far more regularity than in the case of metamorphic rocks. The resulting trellis stream pattern is formed in areas of folded rocks such as the Appalachians. Here the rivers flowing on the weaker layers have cut their valleys much more rapidly than other streams and

thus developed the patterns seen today, where alternating elongated parallel ridges and valleys predominate.

A road map of Pennsylvania reveals a great deal about the topography

DENDRITIC AND TRELLIS DRAINAGE PATTERNS, WEST VIRGINIA. The trellis
pattern is found only in the easternmost part of the state. These same
patterns are present directly north in Pennsylvania.

of the region. The roads which follow the river valleys emphasize the
northeast-southwest alignment of weak and resistant rocks as they appear
in the valleys and ridges of the state. The junction of the resulting trellis
pattern in the central part of the state with the dendritic pattern in the
western part is dramatically sharp, and with confidence we can draw a
line which divides the folded rocks on the east from the Appalachian

Plateau province on the west. The eastern margin of the folded area is only slightly less obvious.

Two other river patterns, the radial and the parallel, are of local im-

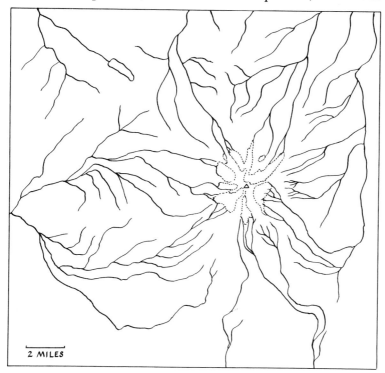

RADIAL DRAINAGE PATTERN, MT. HOOD, OREGON

portance. They are somewhat exceptional in that the pattern is not due to differences in rock resistance but to the initial slope of a relatively new land area, such as the sides of a volcano, a tilted block of the earth's crust, or the slope of a coastal plain. The radial pattern is formed where streams radiate from a central point, and generally indicates a volcanic peak. This pattern is especially noteworthy on a detailed drainage map of the

Hawaiian Islands; it can also be noted on detailed maps of the Cascade Mountains.

An excellent example of a series of parallel streams is supplied by the

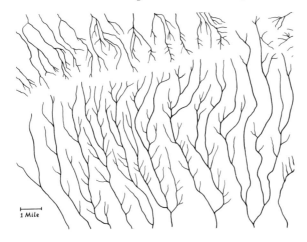

1 Mile

PARALLEL
DRAINAGE PATTERN,
MESA VERDE, COLORADO

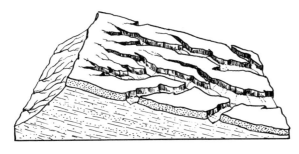

GENERALIZED
BLOCK DIAGRAM
OF MESA VERDE

group of eastward flowing rivers which drain the Coastal Plain from New Jersey to Georgia. The recently drowned seaward ends of the Delaware, Susquehanna, and Potomac, are parallel to such rivers as the Rappahannock, James, Roanoke, and Savannah. These streams owe their parallelism to the fact that they are running down the slope of an uplifted part of the Continental Shelf. Individually, however, they are excellent examples of the dendritic pattern, with tributaries developed at random, essentially uncontrolled by differences in rock hardness.

The Mesa Verde tilted block in southeastern Colorado has a series of streams flowing southward down the dip slope; when considered individually, the major streams with their tributaries form dendritic patterns. Taken collectively, these streams show also a parallel pattern, which is due in this case to the tilted orientation of the resistant cap rock of this part of the Colorado Plateau. It should be emphasized that in the parallel pattern all the major streams flow in the same direction, unlike the trellis pattern where the major valleys are not drained by rivers flowing in the same direction.

Commonly a mature stream is not left in peace to pursue its wandering course across its flood plain for very long. Some cause, such as uplift of the land or an increased flow of water following a change of climate, may rejuvenate it and start it cutting downward again as it did in its early history. Such a rejuvenation will cause the stream to entrench itself in its own flood plain deposits, and as a result sometimes only traces of the former flood plain may be left as terraces on the sides of the valley. If the process of rejuvenation is repeated a number of times a series of terraces will result. Such flood plain terraces will consist of layered sand and gravel, whose structure can be noted in any road cut. Terraces are a common feature along many mature stream valleys.

With continued downcutting a meandering stream may go right through its flood plain deposits and entrench itself in the underlying bedrock. The meandering path of such an entrenched stream is a relic of its former life when it was a mature stream and meandered over a wide valley floor. It explains what might seem at first sight to be an impossible course for any stream to have picked. Excellent examples of such entrenched meanders can be found in many places. The San Juan River in southern Utah has produced some really incredible sinuous cuts thousands of feet deep in the Colorado Plateau. The meandering here was so tight and the neck of land between neighboring stretches of the river so narrow that they have been given the name Goosenecks of the San Juan. A number of the meandering major rivers in the Appalachian Mountains incised their channels following the most recent uplift of the area.

Individual streams produce their own local scenic effects—valleys, rapids, falls, potholes, deltas, fans, levees, and sand bars. Where groups of streams are considered and their over-all pattern can be appreciated, explanations must be in terms of the original slope of the land or the structure of the rocks.

Again we must think in time-dimensional terms. Running water, which is so largely responsible for the scenery which diversifies our drive across the continent, owes its effective force to two things, the sun's energy, which evaporates water from both lands and ocean and raises it high into the air, and the deep-seated diastrophic forces of the past which have raised the land so that rain is given a slope down which to run. What we see now is all part of the eternal cycle—sands and muds washed down by rivers into the sea, these sediments hardened into rocks and raised again to be attacked by running water.

CHAPTER 9

MOST agents of erosion, as we have seen, perform their functions in an obvious manner. Once our attention is called to the various methods of rock weathering and to the way the broken material is carried away, the part played by stream erosion, for instance, in creating landscapes becomes relatively clear. It is a little more difficult, however, to follow water underground in imagination and to picture how and where it travels and

Water under the Land

in what manner it does its job of destruction and deposition under the surface of the earth.

Water moving under the land does a tremendous amount of geological work. Near the surface it plays an important part in the formation of soils, and at somewhat greater depths it helps in the consolidation of sedimentary rocks, as it moves material in solution from one place to another.

The passage of water through rocks is very devious and slow. It follows joint cracks or percolates between mineral grains and rock particles in sand and gravel. Even after these coarser sediments have been turned into sandstones and conglomerates, connected openings adequate for the passage of water may remain open. The finer grained mud and its consolidated equivalent, shale, on the other hand, do not afford openings large enough for the passage of water.

Deep in the fractured zones of the crust, ground water is essentially stagnant. From this condition its motion may vary, perhaps up to fifty feet per day in some very permeable sandstones near the surface. It is actually proper to speak of true underground rivers in only a few cases, generally in connection with the flow of water through the wide openings in cavernous limestone. Elsewhere we may get the false impression of having hit an underground stream when, for instance, a well penetrates a mass of very porous water-soaked material such as gravel or sand and the water pours into the well opening almost as fast as it can be pumped.

The ultimate source of all fresh water is the sea. Water initially in the ocean is evaporated and then precipitated over the land in the form of rain, hail, or snow. After landing, this moisture may be evaporated, flow off on the surface, or sink into the ground. However, nearly all that goes underground eventually emerges at a lower level to join the surface runoff and be subject to evaporation.

Rain percolates into the earth slowly where the openings are tortuous and poorly connected, and more rapidly where they are large and more continuous. Openings do not exist much below 5,000 feet. Thus ground-water circulation is confined to approximately the top mile of the crust. All available cracks and openings in this zone under the larger perennial rivers and lakes are filled right up to the surface. Under the land between such rivers and lakes the depth to the saturated rock will vary. This level of saturation, which is called the "water table" is nearer the surface of the earth after a time of rain, and lower after a period of drought.

The shape of the water table is, in general, a subdued replica of the

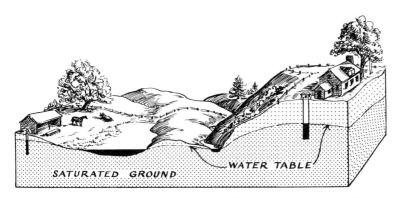

IDEALIZED BLOCK DIAGRAM SHOWING THE RELATIONSHIP OF THE WATER TABLE TO THE GROUND SURFACE

ground surface. It rises under the hills and emerges to coincide with the surface of the perennial lakes and rivers. We can thus imagine hills of water under the land, hills which are of course constantly tending to collapse by slow leakage through the rocks, but which are constantly replenished at the top by new additions of rain water. Ground water emerges as springs and hillside seepages to feed lakes and rivers, which would very soon run dry were it not for their nourishment from such underground sources. Actually, at any one time there is probably very much more fresh water underground than on the surface in all the rivers and lakes combined.

One of the major problems connected with a house in the country is an adequate water supply. In most cases this involves the location and sinking of a well. If a good supply of water is obtained we may be sure that two basic conditions have been met; the bottom of the well is below the water table, and it has penetrated some porous material through which water can flow freely. Such water-bearing rocks are called aquifers.

Unconsolidated materials such as sands and gravel make excellent aquifers. Many consolidated rocks make good water sources too, if they are permeable. Granite, if it is well-jointed, can supply water, and so also can volcanic flows.

The Dakota sandstone which underlies much of North and South Dakota, Nebraska, and adjacent states is a classic example of an aquifer. This water-saturated layer is generally less than 100 feet thick and in many places underlies the surface at a depth of several thousand feet. The layer is upturned and emerges at the surface along the edge of the Rockies. Rain enters it here and percolates down and out under the plains, where thousands of wells have been dug into it. Many of the windmills which break the monotony of the Dakota and Nebraska landscape are pumping water out of this underground source.

Some of the most productive springs in the United States are along a fifty mile stretch of the Snake River in Idaho. The river here has cut down through some very porous lava flows of the Snake River Plain and water gushes out at a tremendous rate. Going across this lava plain we may look on it as a barren stretch of land, but under the surface there is moving through the various openings in the rock a great deal of water, which forms a vast reservoir. Big and Little Lost Rivers disappear into the Plain on the north side, lost in the very permeable lava flows.

Erosion by underground water produces interesting scenic effects on the surface primarily in limestone areas, where it is not unusual to find an almost total absence of surface streams, even though the rainfall may be heavy. What streams there are disappear into the ground, reappearing elsewhere at a lower level, where they may flow for short distances on the surface only to disappear again.

As quickly as rain lands in such a region it sinks out of sight into underground passageways dissolved out of the limestone by previous rain water seeking a lower level through and into any openings it found available. It is in such country that many caves are found, where large pockets of the limestone have locally been removed by the slow steady drip and flow of water as it has percolated through the rocks. Here also the surface of the land may be indented with many hollows of varying size, called "sink holes," formed by the collapse of the surface following subterranean removal of material. Sink holes, disappearing streams, and the absence of any integrated network of surface drainage are the major scenic characteristics associated with the erosion of limestone by subsurface water.

As seen from the air parts of Southern Indiana have a most remarkable appearance. The land of small rectangular fields appears to be literally peppered with dots. Each one is a small sink-hole. In places there are over one thousand in a square mile, with perhaps a total of 300,000 altogether. Large tracts of this country lack surface streams completely.

The type locality for such scenery is in the Karst region of Yugoslavia, along the Adriatic Coast; thus all areas which show this type of scenic pattern are called karst areas. Besides Indiana there are a number of notable karst regions in the United States. Two of them are in the Appalachians— one in the Great Valley of Virginia and Tennessee and the other in an area extending from north-central Tennessee up into central Kentucky. The northern part of Florida is also a karst region.

One of the most interesting features produced by ground water is the limestone cavern. Each of the many caves in the United States demon-

LEHMAN CAVES NATIONAL MONUMENT, NEVADA. *Photo by Devereux Butcher*

strates the seemingly infinite number of variants which nature can make on a few fundamental themes. Luray Caverns in Virginia, Mammoth Cave in Kentucky, and the Carlsbad Caverns in New Mexico are known throughout the country. They and many others, large and small, have been made quite accessible and are well lighted, and they are therefore extremely convenient and enjoyable for the tourist to see.

An excellent example of one of the smaller caves is Lehman Caves of Nevada, now set aside as a National Monument. The entrance lies part way up the west side of one of the large Basin and Range valleys. The rooms in the cavern series are not as large as many in other notable caves but the dripstone shapes are unusually varied and dramatically lighted.

The trip through these caves takes about one hour. The passageways vary abruptly in both size and shape—which is true of all underground openings in limestone. Water flowing through the rock follows the joint systems, but since no mass of limestone is perfectly uniform in solubility, it will be eaten away in detail in a very irregular fashion. The Lehman Caves, since their initial formation, have been slowly filled in by the precipitation of calcium carbonate, in the form of dripstone. And it is this filling which gives the almost infinite variety to such underground scenes.

It is amazing what a diversity of forms can be produced by the simple act of the precipitation of limestone. There are the pointed stalactites hanging from the ceiling, the more stumpy stalagmites growing from the floor, queer platter-shaped formations, complicated pleated draperies, columns of various shape, size, and arrangement, and translucent bacon-striped sheets. And all these airy forms are made of solid resistant rock. When tapped with a hammer, many of the columns produce a strong vibrant tone, and a series of such columns can often be found which vary enough in pitch for a tune to be played on them. With a little study of such a scene, it becomes apparent that there are only two basic scenic forms, which by their modification and combination produce most of the variety to be seen—stalactites and stalagmites. Very commonly a stalagmite will be found growing upward directly underneath a downward growing stalactite, obviously having been built by material brought by the same dripping water which constructed the ceiling stalactite. These two

LEHMAN CAVES NATIONAL MONUMENT, NEVADA. *Photo by Devereux Butcher*

types of growth may eventually join to form a column, and a whole sequence of columns may form a fluted wall. A row of stalactites often follows a crack in the ceiling from which water has oozed and dripped from
many different places.

Pure calcium carbonate is colorless or white, and the more common buff
color of the dripstone deposits, as well as the reds and browns of the
bacon-striped pieces are due to small amounts of impurities deposited with
the lime. Precipitation of this material occurs when there is evaporation,
or a change in temperature or pressure.

Such a cave scene is so infinite in its variety that we may wonder how
conditions can vary so much as to produce such a haphazard arrangement
of shapes and colors. Fundamentally, the diversity must be due to the inter-
play of the complex conditions of solution and deposition combined with
the location and shapes of the openings through which ground water
enters the cave.

The observant traveler may note that, while it is obvious that the cave
he is visiting has been excavated by ground water, now it is being filled in
by lime deposits. And he may ask, why this deposition after erosion? It
has been suggested that caves were formed in the first place by ground
water circulating below the water table and that deposits started to fill
them after the water table had sunk below the cave level.

The solution and deposition of material by water moving through the
earth occurs everywhere, and in places the result may be the formation
of concretions. These hard, oddly shaped objects are found in many sedimentary layers. Their origin is similar to that of petrified wood. The flint
concretions commonly found in chalk layers are built by water seeping
through the chalk and in the process picking up rare bits of silica which
it transports elsewhere to be deposited as growing globules of hard flint.

Such dark gray or black concretions vary in size from a fraction of an inch to a foot or more. Local concentrations of lime or iron minerals may occasionally be found in shales as concretions. At places a fossil or some bit of organic matter has apparently acted as a nucleus around which such minerals have been precipitated.

The form of many concretions may resemble man-made objects or living organisms. Many such features have been seriously taken to be true fossils. It should be stressed, however, that the shapes of concretions as well as other shapes in inorganic nature are to be explained as either the chance deposition or erosion of material.

Of the many odd shapes in nature the naturally produced bridge is one of the most interesting. We have already discussed a number of environments in which natural bridges can be formed. Weathering produces such natural arches as those at the Arches Monument in Utah, and stream erosion can form such features as Rainbow Bridge, also in Utah. The erosive work of ocean waves can carve sea arches, such as Percé Rock in the Gaspé region of Canada, or the arches off the California coast which we have mentioned. The partial collapse of the roof of a lava tunnel may leave a small bridge. Lastly, we find some very notable natural bridges created by ground water in limestone areas.

The limestone bridge in Virginia is well known. U.S. Route 11 at present uses it to cross Cedar Creek. The bridge lies 150 feet above the creek bed, is about 100 feet long, 40 to 50 feet thick, and varies in width from 50 to 150 feet. It is a limestone remnant left after the rock on either side and underneath was removed in solution. It represents the uncollapsed remnant of a roof over a former subterranean river channel. It has been postulated that at one time Cedar Creek flowed on the surface in a slightly circuitous path and that a shortened underground cutoff was developed through enlarged joint cracks. Eventually the cutoff became the main channel, and the subsequent collapse of most of the roof of this channel, a relatively recent event, produced this scenic feature of today.

Perhaps the most dramatic contact that one can make with the world of underground water is in a region of geysers. Practically all the water emitted from such natural fountains is ground water which started as rain and percolated below the surface only to reappear as heated water, sometimes boiling hot.

The notable geyser fields of the world are in Iceland. New Zealand, and Yellowstone Park, in Wyoming. This park probably draws more visitors every year than any other national park in the United States. Yellowstone Lake, the Falls, and the unspoiled areas of forest and meadowland with bears and other wild life have an appeal for most tourists, but there is no doubt that the greatest attraction for the average person is Old Faithful and the geyser basins. The eruptions are spectacular, the landscape produced by the mineral deposits is extraordinary, and there is a slight element of hazard which adds to the experience of strolling about on board walks over the crust—one is reminded that the crust is thin, and that walking on it is extremely dangerous, that "a fall into a boiling pool is fatal."

The mechanics of a geyser are relatively simple. It is merely a periodically eruptive hot spring which emerges from a rather special contorted system of passageways which connect the surface of the land and its supply of rain water with hot rocks at depth. Water filling such a system is heated at its lower extremity. Because of the narrow passageways this heated water does not rise by convection but gets hotter and hotter, and, due to the weight of the overlying cooler water, the water at depth can be heated well above the ordinary boiling point before being turned into vapor. When, however, with continued heating, vapor is produced at this depth, the whole column of water is lifted as a froth by the rising, rapidly expanding superheated steam, and the geyser erupts. Eruption continues until the superheated steam has all escaped, when the passageways are again filled with cool water. Then the whole cycle of heating and eruption starts over again.

Geysers are more or less periodic, but the supply of heat or the supply of ground water varies slightly so that even such geysers as Old Faithful may show some variability of timing. They go into their act when they are ready, when the hydraulics of the situation demand it. As the ranger at Old Faithful likes to point out: "Geysers are always on time for their own needs, they are never late and never early."

The "paintpots" are an intriguing part of the Yellowstone scene. They are bubbling baths of variegated muds showing purples, reds, yellows, and other colors. The colors come from iron and manganese oxides and other impurities in the mud and result from the breakdown of volcanic rock altered by ascending hot waters, which carry the resulting clay to the surface. Some of the hot water may be magmatic, that is, coming directly from the cooling magma which supplies the heat to the area. However, most of the water appearing as hot springs and geysers is undoubtedly reemerging rain water.

Many of the springs in the Park are extremely hot. Every year at Beryl Spring, there are visitors who try the old trick of filling a string bag at the end of a stick with eggs and holding them in the boiling water. In any event, this spring acts as a demonstration of the latent power of the earth, where the imprisoned magma can heat water to the boiling point.

All ground water picks up some material in its passage through the rocks. Well water from a limestone area is "hard" because it carries a

GRAND GEYSER IN THE UPPER GEYSER BASIN, YELLOWSTONE NATIONAL PARK, WYO-
MING. Eruption occurs every 20 to 80 hours and lasts for about half an hour. The
cloud of steam and hot water reaches a height of 180 to 200 feet.

Photo by Josef Muench

large amount of calcium carbonate in solution; cooking kettles quickly build up a thick coating of lime. Hot springs and geysers are especially rich in dissolved material and may contain a large variety of chemical compounds in solution.

At Mammoth Hot Springs in Yellowstone Park the hot water has brought to the surface quantities of white mineral deposits which form high terraces over which the water now flows. And around the vents of some of the geysers a funnel has been built up of these deposits.

The main geysers and hot springs are located in definite zones, but it is possible to find many small ones scattered around the fringes of the larger basins. Small patches of steam and above all the smell of sulphur lead one to them. The smaller hot springs have not been able to take over the land so completely to transform the woods into desolate areas, as have the larger ones.

Geysers and hot springs are of course indicative of the presence of magma under the surface, and, in passing, it is interesting to remember the many other evidences of volcanic activity here in the Park, such as the wall of black glass at Obsidian Cliff and the layer of lava at Tower Junction with jointing so clearly defined as to suggest carefully carved columns.

CROSS-BEDDING, CANYON DE CHELLY, ARIZONA   *Photo by Dorothy Abbe*

CHAPTER 10

FIFTY million years ago small fish swam in the water of a large lake, died, and were buried in the soft muck at the bottom. Today, in the Green River Shale of Wyoming we can trace their outlines, each bone delicately preserved in carbon. Almost two hundred million years ago small, two-legged dinosaurs left tracks like those of out-size turkeys in the soft mud of a river bank and now we find them hardened in rock in the Connecti-

Time and Change

cut River Valley. Five hundred million years ago, tiny trilobites, looking rather like tailless horseshoe crabs, crawled on the ocean bed; today we can dig their fossil remains out of sandstones in the Rockies.

How can geologists talk so glibly about such vast expanses of time? How do they know that the earth is probably well over three billion years old? Moreover, how do they know that one rock is older than another?

Before the present century and the development of the radioactive method of age determination scientists could only give the relative ages of rocks.

There are three basic relationships by which the relative ages of different rock units can be determined without knowing anything about the actual ages. Any igneous rock which intrudes another rock is obviously younger than that which it intrudes. For instance, dikes, volcanic necks, and masses of granite became solid after the formation of the material in which they are found. Secondly, in a sequence of sediments those which are on top are younger than those which are underneath. This is certainly an obvious relationship, but one by the use of which the third relationship, that concerning fossils, was established. No assemblage of life forms, either plant or animal, has ever been repeated in earth history. This means that fossils found in sedimentary rocks can be used to give relative dates, once the sequence of life forms has been worked out.

By the use of these relationships geologists have been able to construct a time chart. The recent eras in earth history, the Paleozoic, Mesozoic, and Cenozoic are distinguished on the basis of the fossils found in rocks, and refer to "old," "middle," and "recent" types of life. For eras before the Paleozoic, in Pre-Cambrian time, fossils are not common enough to be used for dating, so the timing of geologic events is much more tentative. The subdivision of the eras into smaller periods has been accomplished largely on the basis of local physical events with the help of a detailed analysis of the fossils.

When this time chart was developed it was impossible to determine the actual length in years of the various periods, although it was appreciated that geologic time had to be measured in thousands and perhaps millions of years, judging by the obviously slow rates at which some of the features

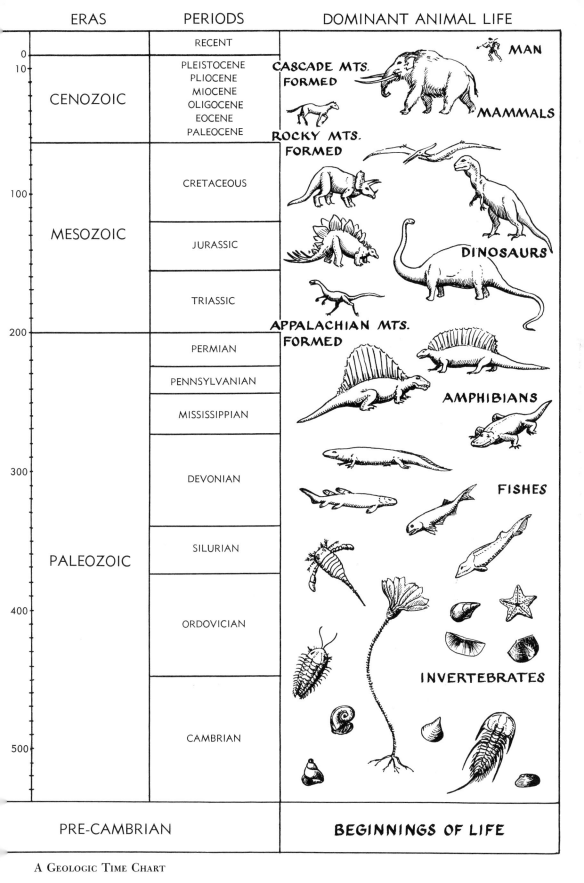

ERAS	PERIODS	DOMINANT ANIMAL LIFE

	RECENT	
CENOZOIC	PLEISTOCENE	CASCADE MTS. FORMED
	PLIOCENE	
	MIOCENE	
	OLIGOCENE	
	EOCENE	
	PALEOCENE	ROCKY MTS. FORMED

MAN

MAMMALS

MESOZOIC	CRETACEOUS	
	JURASSIC	DINOSAURS
	TRIASSIC	

PALEOZOIC	PERMIAN	APPALACHIAN MTS. FORMED
	PENNSYLVANIAN	
	MISSISSIPPIAN	AMPHIBIANS
	DEVONIAN	
	SILURIAN	FISHES
	ORDOVICIAN	
	CAMBRIAN	INVERTEBRATES

Scale markings: 0, 10, 100, 200, 300, 400, 500

PRE-CAMBRIAN — BEGINNINGS OF LIFE

A GEOLOGIC TIME CHART

FOSSIL RIPPLE MARKS IN SANDSTONE, BARABOO, WISCONSIN. *Photo by John Shimer*

of the earth have been formed. For instance, if we assume that geologic events occurred in the past at roughly the same rate as at present it would have taken many thousands of years for the Mississippi River to have built its delta to its present size.

Knowledge of the actual ages of rocks had to wait until the development of the radioactive methods of age determination. The phenomenon of radioactivity was discovered by Henri Becquerel in 1896. Now, techniques have been devised by which the age in years of any mineral containing one of the radioactive elements, such as uranium or thorium, can be found. These elements, which disintegrate spontaneously, change into lead after discarding parts of their masses in a series of steps. This process is very slow. In the case of uranium it takes 7,600,000,000 years for one half of a supply to disintegrate into lead. The age of a rock which has a uranium-bearing mineral can thus be determined by comparing the mass of uranium present with the amount of lead. The longer the mineral has existed the greater will be the proportion of lead to uranium.

For the dating of much more recent geologic events another radioactive element has been found to be of value. This is a variety of the element carbon, carbon fourteen, which is produced from nitrogen by the action of cosmic rays in the upper atmosphere.

Some carbon fourteen, in a constant proportion, is mixed with the ordinary carbon found in all organisms, both animal and plant, while they are alive. When an organism dies and is buried, no more carbon fourteen, or any other carbon for that matter, is added. What carbon fourteen there is breaks down into nitrogen at the fixed rate of one half of the amount present every 5,500 years. Thus the amount of this special form of carbon which is left in organisms and other objects containing carbon is in inverse measure to the duration of burial. In this way the date of burial of anything that once was alive and hence contains carbon, such as wood or objects made of wood, carbonate shells, or animal remains, can be determined if the time of burial has not been more than a few tens of thousands of years. This method of dating is especially useful in archaeological studies and in studies concerning the events at the time of the last glacial advance and retreat, which, as we have seen, was so important in the making of much of our modern scenery.

By using radioactive methods of age determination scientists have found that the earth must be over three billion years old. Throughout this enormous stretch of years the crust has gone through countless cycles of de-

RECENT RIPPLE MARKS ON COHASSET BEACH, MASSACHUSETTS. *Photo by John Shimer*

struction followed by uplift. Parts of the land have been raised into mountains and plateaus, which through milleniums have been gradually washed back into the sea. Continents have been invaded by oceans, and generations of now vanished creatures of the sea have lived and died and have been buried on beaches now perhaps elevated into mountains. In many places we can now find ripple marks preserved in solid mud or sand layers. They were formed millions of years ago on former sea beaches, in identically the same way that similar ones are being made today on innumerable contemporary beaches.

The scenic features we see around us, hills and valleys, lakes, water-falls, and cliffs, are all geologically recent in origin, yet the location and shapes of many of them were determined long ago.

There are many striking features which dramatically reveal the dependence of modern scenery on events of the past. For instance, Tuscorara Mountain in central Pennsylvania, was eroded into its present shape in recent geologic time, but its form and location were determined by the deposition of a layer of sandstone in the sea that filled the great Appalachian geosyncline 400 million years ago, and the folding of this layer 200 million years later. The granite of the Black Hills in South Dakota was made in Pre-Cambrian time over 500 million years ago, and recent erosion has left the odd vertical pillars of the Needles. The location and occurrence of the mesas, buttes, and steep cliffs of the Colorado Plateau and the hogbacks in the folded Rocky Mountains were determined initially by events of the geologic past. Layers of sediments had to be deposited, which on burial were consolidated into alternating layers of resistant and nonresistant rocks. Following this, uplift, as in the Colorado Plateau, or uplift with folding, as in the Rockies, was necessary before erosion could uncover the buried layers and in the process produce the cliffs and ridges we see today.

It is not only hints of past geologic history that we may note; in our travels we may guess at something of the future. We see that the same forces of destruction and rebirth that have resulted in the scenery of today are at work creating that of tomorrow. As soon as a young volcano such as Paricutin becomes extinct, for example, weathering and erosion start to destroy it. The perfect cone with a crater at the top will pass through various stages as it is destroyed. We feel sure that after the crater has been breached at the top and river valleys have been developed on the flanks of the cone, eventually the volcano will be entirely removed and that perhaps only the more resistant material filling its neck will stand exposed to view, much as Ship Rock, in New Mexico, stands starkly on a flat desert.

The landscape which we see today is thus but a momentary scene in the stretch of geologic time. As the world of the past evolved into that of the present, so the world of today, which carries traces of its past history, changes bit by bit into that of tomorrow. The only constant factor in nature is change.

Glossary

AA (ä-ä) A dark-colored, jagged clinkery lava, occurring in angular broken blocks formed by the cracking of a hardened lava crust by molten rock running underneath. Compare with *pahoehoe*.

ALLUVIAL FAN A river deposit having a fan-shaped outline. Built by a mountain stream when it loses velocity at the foot of a steep slope. The land counterpart of a delta.

ANTICLINE An upfold or arch in lay-ered rocks. The opposite of a syncline, with which it usually appears in alternation.

AQUIFER A water-bearing layer of earth, gravel, or rock.

ARÊTE A sharp, jagged mountain ridge between glaciated valleys.

BADLAND A region consisting of a maze of deep gullies, with intervening sharp ridges and pinnacles. Erosion is so rapid and the slopes so steep that vegetation has not been able to take hold.

BASALT Dark-colored, fine-grained igneous rock. Usually occurs in lava flows.

BOMBS, VOLCANIC Spindle or tear-shaped masses of porous lava varying in length from a few inches to a foot or more. Formed from liquid gouts of magma blown out of a volcano and hardened in flight.

BUTTE A term commonly used in the West and Southwest for conspicuous steep-sided isolated hills. Most commonly an erosional remnant in a plateau area; applied also to various other features, such as isolated cinder cones.

CALDERA A large basin-shaped depression at the top of a volcano many times larger than the average crater.

CINDER CONE A small volcano constructed primarily of ash and cinders. See also *strato volcano; shield volcano.*

CIRQUE A bowl-like amphitheater cut into the side of a mountain by glacial erosion. Formed at the head of a valley glacier.

COASTAL PLAIN A plain, underlain by horizontal or gently sloping sediments, bordering the ocean. It generally represents a recently emerged section of the sea floor.

COLUMNAR JOINTING Parallel rock fractures, usually vertical, associated with basalt lava flows or sills. Such jointing results from the shrinkage of lava on cooling.

CONCRETION An irregular nodular  or disc-shaped body found in sedimentary rocks. Formed by local concentration of cementing materials, such as silica, calcite, or iron oxide, some concretions measure a foot or more across.

CONGLOMERATE A sedimentary rock formed of more or less rounded pebbles cemented together.

CONTINENTAL GLACIER An ice sheet which covers an appreciable part of a continent, overriding hills and valleys alike. Continental glaciers exist today in Greenland and Antarctica. Compare with *valley glacier*.

CROSS-BEDDING A structure present  in many sandstones where the layering in the individual beds forms an angle with the principal bedding planes.

CUESTA An asymmetric ridge sloping  steeply on one side and very gently on the other, capped by a resistant layer.

DECOMPOSITION The chemical breakdown of the minerals in a rock. Synonymous with chemical weathering.

DELTA The deposit formed by a stream  as it enters a body of water and drops its load of sediments. Many deltas are roughly triangular in plan resembling the Greek letter delta with the apex pointing upstream.

DIASTROPHISM The process by which the crust of the earth is deformed. It includes folding and faulting, and is one of the three major processes which affect the earth's crust. The other two processes are igneous activity and gradation (weathering and erosion).

DIKE A tabular body of igneous rock  which cuts across the structure of preexisting rocks into which it has been intruded.

DISINTEGRATION The mechanical, as opposed to the chemical, breakup of rocks under weathering conditions. It results from such factors as frost action, the wedging effect of plant roots, and changes in temperature.

DRIFT A general term for all glacial deposits, whether deposited directly from the ice or by meltwater.

DRIPSTONE Any calcium carbonate deposit formed in a cave. See also *stalactite* and *stalagmite.*

DROWNED COAST Characterized by many estuaries and islands; results from the sinking of land relative to the sea and the subsequent inundation of coastal areas.

DRUMLIN An elongated hill composed of till deposited by a continental glacier. The longer dimension varies generally from one quarter to one half mile, and the height from 50 to 100 feet.

EROSION The wearing away of the land by any one or a combination of the following five agents; streams, glaciers, wind, underground water, or ocean waves and currents.

ERRATIC BOULDER A boulder transported and dumped by a glacier.

ESKER Associated with glaciation, a  sinuous ridge, 10 to 100 feet high, composed of roughly sorted sand and gravel; often resembles a railroad embankment.

ESTUARY A tidal inlet by which a stream enters the sea. Formed as the result of a rise in sea level and the consequent drowning of the seaward end of a former stream valley.

EXFOLIATION The peeling off of layers of rock. It produces round domal rock forms.

EXTRUSIVE Applied to an igneous rock consisting of lava which has flowed out at the surface of the earth.

FAULT A fracture in a rock along
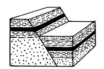 which adjacent
blocks have shift-
ed with respect
to each other.

FAULT BLOCK MOUNTAIN A mountain bounded on one or both sides by
faults. Examples are the Sierra Nevada,
the Grand Tetons, and many ranges in
the Basin and Range region of the
United States.

FIORD A deep, narrow, steep-walled
inlet of the sea; the result of the partial
submergence of a glaciated valley.

FLATIRON A portion of a hogback
 ridge eroded into
a triangular shape;
occurs in series
on the flanks
of mountains.

FLOODPLAIN The flat area adjoining
a mature or old stream; may become
flooded at times of high water.

FUMAROLE A vent in a volcanic area
emitting various gases and fumes.

GEOSYNCLINE A large area on the
earth's crust which has slowly subsided
throughout long periods of time and in
which thick layers of sediments have

been deposited, often to a depth of
many thousands of feet.

GEYSER A periodically eruptive hot
spring.

GRADATION The process of weathering and erosion by which the earth's
surface is worn down. This is one of the
three major processes affecting the
earth's surface. The other two are igneous activity and diastrophism.

GRANITE A coarse-grained, light-colored intrusive igneous rock.

HANGING VALLEY A valley tributary
to a U-shaped glacial trough. Left "hanging" as the result of the rapid deepening
and widening of the principal valley by
glacial erosion.

HOGBACK A ridge formed of a resist-
 ant, steeply dip-
ping layer be-
tween less resist-
ant material. It
is an erosional remnant resulting from
the removal of material from either side.

IGNEOUS ACTIVITY (VULCANISM)
One of the three major processes affecting the earth's surface. This includes the
production of lava flows, volcanoes,
sills, and dikes.

IGNEOUS ROCK A rock formed by the solidification of either magma or lava.

INCISED (OR ENTRENCHED) MEANDERS

Formed when a rejuvenated meandering stream cuts a deep valley for itself through floodplain material and into underlying bedrock. The path of such a stream is inherited from the former meandering path.

INTRUSIVE A body of igneous rock formed beneath the surface of the earth by the solidification of a mass of magma which has penetrated into or between other rocks.

JOINT A fracture or crack in a rock; generally occurs as one of a set of more or less parallel cracks. A joint differs from a fault in that there is no relative motion of the rock masses on either side of the fracture.

KAME A small conical hill composed of roughly sorted sand and gravel, deposited by melt waters from a glacier in a hole in a mass of stagnating ice or as a very steep fan-shaped deposit at the edge of the ice.

KAME TERRACE A terrace deposit of

stratified sand and gravel laid down between a mass of stagnating ice and a valley wall.

KARST TOPOGRAPHY A type of landscape developed in a limestone country, consisting of many sink holes and disappearing streams and caves, where the drainage is primarily underground.

KETTLE HOLE A bowl-shaped depres-

sion, usually from 20 to 50 feet deep and up to a few hundred feet in diameter, found in a glacial deposit. It is produced when a detached block of ice buried in the deposit eventually melts, causing a collapse of the surface.

LAVA Liquid rock as it emerges at the earth's surface from a volcano or fissure. While still underground it is called magma.

LIMESTONE A sedimentary rock composed of calcium carbonate.

LOESS A wind deposit of dust produced by glacial erosion; characterized by its ability to stand in steep cliffs.

MAGMA Liquid rock under the earth's surface. When magma emerges from a fissure or a volcano it is called lava.

MATTERHORN PEAK (HORN) A high pointed peak left as a residual feature by mountain glaciers.

MATURE RIVER VALLEY A valley with a wide flat floor or floodplain veneered with sediments over which a meandering stream flows. When the valley is many times the width of the meander belt it may be considered old.

MEANDER A rounded loop-like bend in the course of a mature or old river.

MESA A flat-topped mountain or table-land, bounded on at least one side by a steep cliff. A partially detached part of a plateau.

METAMORPHIC ROCK A rock formed by physical or chemical changes in some previous igneous or sedimentary rock.

MONADNOCK A hill rising above the level of a peneplain; an erosional remnant.

MORAINE A glacial deposit of till, or unsorted mud, sand, and gravel. Such a deposit may form an extensive ridge which marks the terminus of a glacier (terminal moraine); it may be spread out over the land as a thin widespread deposit (ground moraine); it may show up as a pile of debris riding on or in the middle part of a valley glacier (medial moraine); or it may have been left at the side of a valley glacier (lateral moraine).

OBSIDIAN A volcanic glass, generally black in color.

OUTWASH PLAIN A very gently sloping plain composed of stratified layers of sand and gravel deposited by meltwater from a glacier.

OX-BOW LAKE A crescent-shaped lake on a flood plain, which fills an abandoned meander.

PAHOEHOE (pä-hō-ä-hō-ä) A dark-colored lava showing a billowy or ropy surface. Compare with *aa*.

PEGMATITE A dike-like body of igneous rock composed of very large minerals, which may be many feet across. Most pegmatites are composed primarily of quartz and feldspar with some mica.

PENEPLAIN An extensive nearly flat surface of erosion.

PERMEABILITY The capability of a rock for transmitting a fluid.

PIEDMONT GLACIER A mass of ice at the foot of a mountain formed by the coalescence of several valley glaciers.

PLACER DEPOSIT A deposit of heavy minerals concentrated by running water; found in the lower layers of gravel in a stream channel.

PLAIN A level area of low relief underlain by essentially horizontal rock layers.

PLATEAU A region of horizontal rock layers at a high elevation which have been dissected by deep canyons or river valleys.

PLAYA LAKE A temporary lake found at the center of an undrained desert basin. On drying out such a lake will commonly leave a salt deposit.

PLUG DOME A steep-sided protrusion

of lava in a volcanic crater.

PLUNGE POOL A large pothole formed at the foot of a waterfall.

POROSITY The percentage of open space in a mass of rock or sediments.

POTHOLE A roughly circular hole ground into the bedrock of a stream channel by the abrasive action of swirling sand and gravel.

PUMICE A light-colored, porous lava, very light in weight.

REFRACTION Change in direction of

wave motion. On approaching an irregular coast, ocean waves bend so as to hit the shore more directly along their entire length.

RELIEF As applied to a region it refers to the difference in elevation between the highest and lowest points.

SANDSTONE A sedimentary rock composed of cemented sand grains.

SCHIST A finely foliated metamorphic rock. A common variety is mica schist, in which parallel flakes of mica are predominant.

SCORIA A dark-colored, very porous igneous rock. Occurs only in lava flows.

SEDIMENT Applied to any layer of

loose material, such as mud, sand, or gravel, dumped by one of the erosional agents.

SEDIMENTARY ROCK A rock formed by the solidification or consolidation of sediments. Layering or stratification is the most obvious characteristic of such rocks.

SHALE A sedimentary rock resulting from the consolidation of mud.

SHIELD VOLCANO A broadly convex  volcanic cone, built up almost entirely of rather fluid basaltic lava. Slopes are generally from 2 degrees to at most about 10 degrees. The volcanoes of Hawaii are examples.

SILL A tabular igneous rock body intruded between preexisting sedimentary layers.

SINK HOLE A funnel-shaped depres- sion in a limestone region, resulting from the collapse of the surface into an underground hollow.

SNOWLINE The elevation at which snow exists throughout the year.

SPATTER CONE A low steep-sided volcanic cone built up of spattering gouts of lava emitted spasmodically from a vent.

SPIT A hooked shaped peninsula of  sand built by waves and currents.

STACK An offshore rocky islet. A residual feature isolated from the land by the removal of the surrounding material by wave action.

STALACTITE A cave deposit of cal-  cium carbonate hanging icicle-like from the ceiling.

STALAGMITE A conical, post-like deposit of calcium carbonate growing upward from the floor of a cave.

STRATO VOLCANO A volcano composed of alternating layers of lava and ash. The slopes of the cone are a great deal steeper than those of a shield volcano.

STRIATIONS, GLACIAL Scratches found on rock ledges in glaciated regions. Formed by the abrasive action of boulders and pebbles frozen into the base of moving ice.

SYNCLINE A downfold or trough-like  feature in folded rocks. The opposite of an anticline, with which it usually appears in alternation.

TALUS A slope of broken rock fragments at the base of a cliff.

TARN A small mountain lake, especially that found in the scooped out bottom of a cirque.

TILL A glacial deposit of unsorted material composed of mud, sand, and gravel laid down directly by the melting ice.

TOMBOLO A sand bar which connects  an island with the mainland.

UNCONFORMITY An erosion surface 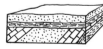 which separates two masses of rock. A common type of unconformity shows an angular relationship between an older tilted or folded sequence which was eroded and a newer sequence of rocks deposited on the erosion surface.

U-SHAPED VALLEY A valley showing a U-shaped cross profile; formed by glacial erosion.

VALLEY GLACIER Sometimes called alpine or mountain glacier. It is a stream of ice confined to a valley and usually starts in a growing cirque.

VOLCANIC NECK An isolated column  or hill of igneous material, representing the solidified filling of the pipe or vent up which lava came to form a volcano. Its presence obviously implies deep erosion.

V-SHAPED VALLEY A valley having a V-shaped cross profile; characteristic of youthful stream valleys.

VULCANISM see *igneous activity*

WATER TABLE The upper surface of the zone of saturated rocks under the earth's surface. It is usually at a higher elevation under hills and lower under valleys.

WEATHERING The decay and break-up of solid rocks at the earth's surface.

YOUNG RIVER VALLEY A valley in its early stages of development; characterized by a V-shaped cross profile. A stream in such a valley may possess falls and rapids, and flows on bedrock for much of its length.

Index